A Geologist Speculates

ON GEMSTONES,
ORIGINS OF GAS AND OIL,
MOONLIKE IMPACT SCARS ON THE EARTH,
THE EMERGENCE OF ANIMALS AND CANCER

John M. Saul

ISBN 978-2-37081-004-5

By the same author:

The Tale Told in All Lands, Les Trois Colonnes, Paris

Envois de manuscrits:

Les 3 Colonnes – 50 rue Saint-Sabin – 75011 Paris

Contents

Preface

During a varied career in geology, I have observed two matters that seem to stand out as fundamental truths. First of all, there is a great deal that we do not understand about the workings of the Earth. But even so, it is clear that everything is connected, not just the many aspects of geology itself — mineralogy, earthquake studies, exploration for oil and gas, and much else — but also astronomy, biology, and even cancer studies and the emergence of complex societies. The connections are not immediately evident in our usual interactions with the world around us, however, or even by plunging into specialized studies, because so many key events took place in the exceedingly distant past, and so much has changed since.

Geology, along with astronomy, turns out to be a good base from which to explore the great nexus of "things connected" because its subject matter includes events that set the stage for much of what was to come.

Chapter 1 of *A Geologist Speculates* deals with the formation of certain transparent gemstones, garnet, ruby, tourmaline, and tanzanite, among them. The elusive conditions that allowed humdrum materials to crystallize in a transparent "gemmy" manner have puzzled scientists at least as far back as the times of Robert Boyle (1627–1691). Hints how such exceptional crystals were able to form include the presence of gases with an "organic" odor associated with many gemstone deposits, their common association with graphite, and the rough alignment of many gemstone occurrences more or less parallel to modern and ancient coastlines.

There is a logical sequence and progression to the chapters, although they can be read separately. Chapters 1 and 4 require somewhat more background than others.

Chapter 2 treats the origin of gas and oil. Oil companies and their exploration teams certainly appear to have done a good job in keeping the world supplied with energy. Is it possible that they have got the science wrong? Some have thought so, including Sir Robert Robinson, Nobel Laureate in chemistry in 1947, who argued that petroleum was a material of inorganic, "mineral" origin to which molecules of biological origin had been added. Robinson wrote of the "duplex origin of petroleum," and it is likely that he was correct, unless the origin was actually "triplex" or "multiplex." For, as suggested by their widespread occurrence throughout the Solar System and beyond, hydrocarbons are easy to make. Earthquakes, volcanism, and the world's reserves of petroleum, and also of water, are all addressed in this chapter.

In Chapter 3, I commence with the scientific argument that the Earth and Moon are neighbors in space and that our planet could not have escaped the great bombardment that scarred the Moon early in the history of the Solar System, leaving the lunar surface saturated with impact scars. These days it is almost universally believed that traces of this bombardment have been eliminated from the surface of our own planet by rain, wind, cycles of erosion and, most of all, by the workings of plate tectonics with the subduction and deep burial of much or the Earth's crust. I argue differently, claiming that the impact scars remain but that scientists have not established the correct criteria for recognizing them. Parts of my argument, which continues into Chapter 4, hinge on the geological occurrences of various transparent gemstones and the locations of certain regions rich in oil and gas.

Much of Chapter 4 deals with the first appearance of large complex animals, what is commonly referred to as the "Cambrian Explosion." This event, or, rather, this episode in the Earth's history, was also marked by the first formation of particular types of gemstone deposits. Excluding special cases treated in Chapter 1, no deposits of garnet, ruby, tourmaline, tanzanite, and numerous other types of transparent gemstones were formed before the approximate time of the Cambrian Explosion. In short, I argue that exceptional mountains — including a "supermountain" — were formed around this time, that deposits of gemstones were formed in their root-zones, and that the existence of complex animals is due to the erosion of these mountains and the materials that were then washed into the seas. For many, however, the main interest of Chapter 4 is the examination of the nature of cancer, a pathology that did not, and could not, exist before the "Explosion." In brief, I argue that the supermountain, the gemstone deposits, the complex animals who are our ancestors, and cancer all came into existence during the same broad interval of geological time, that all were consequences of the workings of plate tectonics, and that the workings of plate tectonics were conditioned and guided by 3-D scars left by the same bombardment that marked the Moon.

In Chapter 4, I evoked Darwinism and the workings of evolution, and in my short final Chapter 5, I consider three life-forms that might arguably have escaped the workings of Darwinism. One case is that of human beings. I show that Darwinism is indeed universal, but I also accept that humans are somehow different from all other creatures, as is almost universally accepted. Darwin did not have the last word. A surprising "post-Darwinian" factor came into play during the first generation of the Paleolithic when religion, purposeful observations of earth and heavens, and mythmaking were all inserted into the ways of humans, inserted virtually instantaneously.

Acknowledgments

This book has been stimulated in part by the work of the Deep Carbon Observatory.

I gratefully acknowledge the help of many individuals in various aspects of this project. For their useful help and observations with Chapter 1, I thank Patrick Aldridge, Richard B. Berg, George Bosshart†, Jim Clanin, Yves Deschamps, I.E. Eliezri, Tony Fallick, Gaston Giuliani, Michael Gray, Roger Harding, Richard W. Hughes, Jaroslav Hyrsl, John Jaszczak, Adrian Jones, Bill Larson, Elizabeth Le Goff, Peter Lyckberg, Alain Mercier, Victor Nechaev, Yasson Papaeliopoulos†, Eckehard Julius Petsch, Mark Saul, Karl Schmetzer, Dietmar Schwarz, William B. "Skip" Simmons, Cedric Simonet, Ted Themelis, and Brian Upton.

I especially thank Walter A. Balmer, Federico Pezzotta, John B. Southard, F.L. Sutherland, and Hanco Zwaan for their careful reviews of this chapter.

In Chapter 2, I thank Vladimir Kutcherov and Steven Soter for their detailed reviews. I also thank John B. Southard for helpful advice and Irina Plotnikova and Vladimir Trofimov for sharing their research results.

I thank Jan Bergström†, William H. "Bill" Byler†, Morvan Derrien, Charlotte Fiquet, Benjamin Foulatier, Janice A. Glaholm, Anièce Lawniczak-Heijnen, Tom McCollom, and Steven Soter for useful observations and generous help in Chapter 3.

I especially thank Vladimir Koutcherov and John B. Southard for detailed reviews of this chapter.

Jere H. Lipps, John B. Southard, and James Trosko provided helpful reviews of Chapter 4.

Part of the allure of gemstones is that they delight the eye. It has been my hope that this volume should succeed to some degree on the same level. I am indebted to all who provided the images shown on these pages. In particular I wish to thank Roberto Appiani, Federico Bärlocher, L.-D. Bayle, Joe Budd, Mia Dixon, A. El Albani, Tom Epaminondas, Ludovic Ferrière, Asia Gajowniczek, J. Gajowniczek, Ruedi Homberger, John A. Jaszczak, Marli Miller, Federico Pezzotta, Albert Russ, Mariya Solovyeva, Ted Themelis, and Robert Weldon for their useful and beautiful photographs and Hildegard Sennac for her fine paintings.

Not all of these friends and colleagues agreed with my ideas and speculations, for which I myself remain fully responsible.

I also acknowledge the contributions of H. Dale Langford to the design of this volume and for his meticulous editing.

I especially thank Jesse H. Ausubel for his encouragement and help in all aspects of this project, and the Richard Lounsbery Foundation for its backing.

Ruby with an overgrowth of non-gem red corundum, Tsavo West National Park, Taita-Taveta District, Kenya. Maximum width across the gemmy section, approximately 1.5 cm. Mined 1974.

1

Gems

Figure 1.1 Garnet, Minas Gerais, Brazil. Collection Alexandre Delerm.

THE ANCIENT MESOPOTAMIANS prized attractive stones, and by 3000 B.C. each of their seven "planets" — which included the Sun and the Moon — had been assigned a "corresponding" gem.[1] The original correspondences are uncertain, but red garnet appears to have been taken as an Earthly counterpart of the red planet Mars, and moonstone, a variety of feldspar, was presumably associated with the Moon (**Figs. 1.1, 1.2**). Each planet was also assigned its color, with aquamarine (**Fig. 1.3**) and blue sapphire (**Fig. 1.4**) perhaps associated with Venus. Lapis lazuli, an opaque stone whose blue is star-speckled with crystals of pyrite (**Fig. 1.5**), represented the entire nighttime sky, and the 6000- or 7000-year-old lapis lazuli workings of Sar-i-Sang in Afghanistan (**Fig. 1.6**) are said to be the world's oldest operating mines.

Each planet also had its metal, herb, animal, insect, snake, water snake, grain, part of the human body, rock, tree, and much else.[2] The rabbit or hare was a representative of the Moon[3] because of the short periods needed to produce a generation of rabbits in the world below or the rebirth of the lunar month in heaven above. In time, correspondences would be extended by Mesopotamians and others to include more subtle aspects of the world, temple dance steps, for example. Much Mesopotamian science survives as cuneiform lists, evidence of a grand multiethnic attempt to inventory the entire sky (by astronomy) and the entire Earth (by "the lesser astronomy," which is to say, alchemy[4]) and then to correlate the two (by astrology). This is where and how science began.

Constellations, stars, and sectors of the zodiac[5] also acquired their counterparts, and in time the proliferation of opinions, translations, and mistranslations would lead to much confusion, an inevitable outcome considering that the archaic correspondences between the World Below and the World Above do not exist outside the human mind. Through it all, an appreciation for transparent, hard, and colored stones endured.

In the intervening centuries and millennia, special value was given to diamond (**Fig. 1.7**). As the hardest material known, it was also the most enduring. Diamond was thus in some manner understood as encapsulating the essence of longevity and immortality. (For mineralogists, "hardness" means resistance to scratching and is measured on a scale

Figure 1.2 Moonstone, Mogok, Burma. 2.0 x 1.5 x 1.5 cm; 38.50 carats (1 carat is equal to 1/5 of a gram). Photo: Mia Dixon/www.PalaMinerals.com

Figure 1.3 Beryl (aquamarine). Poté, near Teófilo Otoni, Minas Gerais, Brazil. 3.9 x 1.9 x 1.9 cm. Collection J. Saul. Photo: ©L.-D. Bayle.

Figure 1.4 Corundum (sapphire), Balangoda, Ratnapura, Sri Lanka. 3.5 x 1.5 cm. Photo: Mia Dixon/ www.PalaMinerals.com

Figure 1.5 Lapis lazuli star-spangled with crystals of pyrite. Approximately 5 cm across. Sar-i-Sang, Badakhshan Province, Afghanistan. Photo: Mia Dixon/www.PalaMinerals.com

Figure 1.6 On the way to the lapis lazuli diggings, believed to be the world's oldest still-operating mines, Sar-i-Sang, Badakhshan Province, Afghanistan. Painting by Hildegarde Sennac.

Figure 1.7 Trio of diamond crystals showing typical crystal forms. From left to right: 3.44 carats (11.00 x 10.40 x 3.55 mm); 9.04 carats (12.52 x 9.49 x 10.16 mm); and 7.90 carats (13.52 x 10.76 x 5.85 mm). Photo: Mia Dixon/ www.PalaMinerals.com

Figure 1.8 Neolithic (4000–2000 B.C.) jade ax found near Canterbury (Kent) of material quarried in the Alps. Length 21.9 cm. ©Trustees of the British Museum. As is common for such axes, it shows no signs of having been used for cutting or chopping.

of 1 to 10, with the softest mineral, talc, as 1 and the hardest, diamond, as 10.*) Similar ideas adhered to jade, among the toughest of natural materials. Jade, the "ax-stone" of antiquity (**Fig. 1.8**), is about ten times as tough as the best traditional ceramics. From China to Mexico, jade was the "Stone of Immortality." ("Toughness" is defined as resistance to breakage; jade is tough as a consequence of the interlocking of tiny crystals.)

These archaic ideas, from times when science and religion were indistinguishable, are at the origin of our notions of beauty and esthetics. They have led us to prize transparent "crystalline colored gemstones" (CCGs[6]), the formation of which constitutes the subject of this chapter. It turns out that the story is not a simple one.

More than 50 different types of hard and normally opaque minerals with a "stony" allure occasionally crystallize in a transparent "gemmy" manner. These CCG-forming minerals, which amount to approximately 1 percent of all minerals currently known, range from the common — garnet and feldspar, for example — to obscure substances whose names are unfamiliar even to most geologists.†

The compositions of gem-forming minerals are for the most part unexceptional. Most are composed primarily of common elements such as oxygen, silicon, calcium, and aluminum. Terms like "cesian beryl"[7] (**Fig. 1.9**) simply indicate that the particular

* Each mineral has its characteristic resistance to scratching. This is expressed on a scale of 1 to 10 defined by the minerals talc, gypsum, calcite, fluorite, apatite, orthoclase feldspar, quartz, topaz, corundum, diamond.

† A mineral, by definition, is a naturally occurring substance formed by a geological process, and with a definable chemical composition and a specific crystalline structure. The definition these days also requires that it be solid at room temperature with exceptions made for mercury and ice. Rocks are mixtures of minerals. Lapis lazuli and jade are rocks, not minerals.

Figure 1.9 Cesium-rich beryl ("morganite," named after the banker and mineral collector J.P. Morgan), 26 cm across, Brazil. Collection of the Natural History Museum of Milan.
Photo: Roberto Appiani.

Figure 1.10 "Stony" (non-gem) watermelon tourmaline ("elbaite," named after the island of Elba), Dunton Gem Mine, Newry, Maine. Approximately 5 x 5 x 5 cm. Specimen from the collection of the A. E. Seaman Mineral Museum. Photo by John A. Jaszczak.

Figure 1.11 Gem-quality watermelon tourmaline (elbaite), Santa Rosa, Minas Gerais, Brazil. 8 x 6 x 24 mm. Specimen from the collection of the A. E. Seaman Mineral Museum. Photo by John A. Jaszczak.

specimen of the mineral in question (beryl) contains small traces of a rare element (cesium). Such traces (which may themselves be exceptional, even if the main constituents of the mineral are not) may be commercially important; traces of manganese and chromium, for example, are known to produce attractive colors.

Another term, "watermelon tourmaline," for red crystals with a green rind (**Figs. 1.10, 1.11**), is used for a variety of the mineral tourmaline, but neither the red zones (which have a significant content of manganese) nor the green zones (with traces of iron) are difficult to understand in themselves, nor is the zoning. The unexplained feature is the transparency of parts, red and green alike, of certain crystals of watermelon tourmaline.

"Stony," opaque, crystals of watermelon tourmaline have the same chemical composition and crystal structure as transparent specimens. In like manner, stony, opaque crystals of the mineral andalusite have the same chemical composition and crystal structure as transparent gem andalusite (**Figs. 1.12, 1.13**), and so on right through the alphabet to the mineral zoisite (**Figs. 1.14, 1.15,** and **List 1**). *The only known difference between the common stony opaque form of these minerals and their corresponding gem varieties is their transparency, produced by the near-perfection of the crystallization.* The cause of the difference between gems and non-gems is a long-standing puzzle for gemologists and mineralogists alike, none of whom have been able to adequately explain how gems form, where they form, when they form, or why they are rare.

As best as is known, the gem-forming minerals in **List 1** have nothing notable in common other than their very rare occurrences as transparent gems.

Figure 1.12 "Stony" (non-gem) andalusite, Praxmar, Tyrol, Austria, 10 x 8.5 x 8 cm. Collection of the École nationale supérieure des mines de Paris, n°3574. Photo: ©L.-D. Bayle.

Figure 1.13 Gem andalusite, Minas Gerais, Brazil, 20 x 18 x 4 mm. J. Saul collection. Photo: ©L.-D. Bayle.

Figure 1.14 "Stony" (non-gem) zoisite, Uluguru Mountains, east-central Tanzania, length 4.3 cm. J. Saul collection.

Figure 1.15 Gem zoisite (tanzanite), Merelani, Arusha Dist., Tanzania. Size of crystal approximately 1.0 cm. J. Saul collection. Photo ©Robert Weldon, Gemological Institute of America (GIA).

List 1: Durable minerals that may crystallize in a transparent "gemmy" manner

andalusite	grandidierite	quartz**
axinite	hambergite	rhodizite
beryl*	hibonite	rutile
beryllonite	humite	scapolite
bromellite	hypersthene	serendibite
chondrodite	jeremejevite	sillimanite
chrysoberyl*	johachidolite	sinhalite
clinohumite	kornerupine	spinel
cordierite	kyanite	spodumene
corundum	londonite	taaffeite
danburite	magnesioaxinite	topaz
diamond**	manganotantalite	tremolite
diopside	olivine*	minerals of the tourmaline supergroup
epidote	painite	vesuvianite
euclase	petalite	zircon
several types of feldspar	pezzottaite	zoisite
minerals of the garnet group	phenacite	
	pollucite	

* The minerals beryl, chrysoberyl, and olivine are partly excluded from my definition of crystalline colored gemstones, as explained later.
** Diamond and quartz are entirely excluded from my definition of crystalline colored gemstones, as explained later.

Gems are not only rare; they are peculiar

Gems are commonly found in the rocks in which they actually crystallized. Such occurrences are known as "primary" or "hard-rock" deposits[8] and strikingly *different types of primary gem deposits are commonly, even normally, found close to one another*. A century ago, A.W.G. Bleeck concluded his report on the ruby deposits of Burma, writing that it was "difficult to conceive that two or more similar occurrences in one and the same country should be of absolutely different origin."[9] Matters were no clearer eight decades later when Al Levinson suggested that there had been two generations of ruby at Mogok, one formed by broad regional forces and a later event that produced gems at the contacts of different types of rocks.[10] It was the same elsewhere too, Gary Bowerbox and B.E. Chamberlin observing that "the origins of the Afghan gem minerals are as diverse as the rocks in which they are found,"[11] and in Kenya and Tanzania matters are similar.

Mother Nature has numerous ways of producing transparent colored gemstones. A list I once compiled from published texts included 15 different geological recipes for sapphires, but there are surely more. Yet, aside from a very few exceptions individually noted, all the ways by which transparent colored gems are formed fall into one of three major categories:

1. The first category is through *metamorphism*, that is to say, by chemical and physical changes that take place when rocks recrystallize under the influence of high temperatures and pressures while still essentially solid. In these circumstances, various fluids are expelled and diverse chemical reactions take place.*
2. Gems also form in certain very coarse-grained granite-like rocks called "granitic pegmatites," or simply *pegmatites*, some of which contain rare elements.
3. *Magmatic* gems are found in association with particular types of lava. Debate continues whether the gems crystallized within magmas (melted rocks beneath the surface of the Earth, with or without crystals in suspension), or whether they crystallized in other circumstances and were later entrained and brought to the Earth's surface by rising magma.

Figure 1.16 Green tsavorite garnet, Lemshuku, northern Tanzania.
Photo: Swala Gem Traders, Arusha, Tanzania; www.swalagemtraders.com

Crystalline colored gemstones formed in the course of metamorphism

Neither the rock type, nor the mineralogy, nor the chemistry is of evident use in attempting to understand why minerals crystallized as transparent gems in the course of metamorphism.

▸ Gemstones formed in the course of metamorphism are pleasing to the eyes. Not so the nose. In many primary deposits, the host rocks emit a strong organic smell when mined or when trimmed of waste material before faceting. In some cases, a smell emerges from materials that fill microscopic fractures or other inclusions within the gems themselves.

Such odors have been noted in association with zoisite[12] (whose gem variety is known as tanzanite) from Tanzania (**Fig. 1.15**), from green tsavorite garnet (Kenya and Tanzania; **Fig. 1.16**), ruby (Kenya), tourmaline[13] (Brazil; **Figs. 1.11, 1.17**), rubies from impure marble in the Himalayas,[14] and in the Mogok Gem Tract of Burma (Myanmar).[15]

The first, uppermost, and best production of rubies from the John Saul Ruby Mine in Kenya came from the "Main Pit Crumbly Zone" (**Fig. 1.18**), and the exceptional odor

Figure 1.17 Tourmalines, Brazil. Collection Alexandre Delerm.

* When metamorphism is accompanied by significant chemical exchanges, the term *metasomatism* is used.

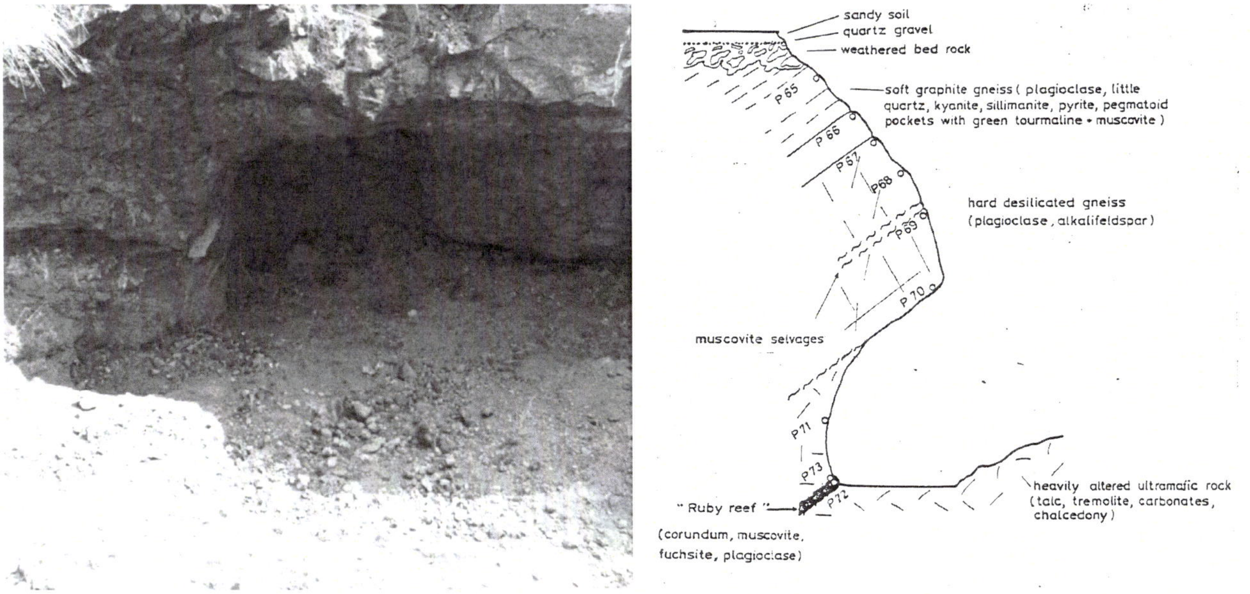

Figure 1.18 (a) Crumbly Zone, Main pit, John Saul Ruby Mine, Tsavo Park West, Taita-Taveta District, Kenya, early 1974. The view is toward the north-northeast. The tall grass in the upper left-hand corner gives an indication of the scale. (b) Geological cross section of the pit in the previous figure looking toward the south-southeast. Austrominerals (1978) p. 102; courtesy of W. Werneck.

emitted during the trimming of these stones was sufficiently strong and disagreeable to cause lapidary workers in the French Jura Mountains to keep their workshop windows open during the European winter; (**Fig. 1.19**). In Burma, where gem mining is also associated with a stench, one offending substance has been identified as skatole,[16] whose name comes from the Greek word for dung. Generalizing, one specialist reported that all the primary metamorphic gemstone deposits he had visited emitted "the bad smell."[17] (Bad-smelling rocks do not necessarily host occurrences of gemstones, but the presence of an organic odor is a usable prospecting tool.)

The liberation of gas from within rough gems in the course of trimming suggests that some or all of the gas had been present from the time the gems crystallized, that it had been trapped both *where and when* the gems crystallized.

Figure 1.19 Saint-Claude, town in the Jura Mountains of France, where windows of a lapidary workshop were kept open year round due to the odors emitted from certain gems while being trimmed.

▸ *Some gemstones occur in rocks that break apart extraordinarily easily.* At the Scorpion tsavorite (green garnet) mine in Kenya, "the best stones could be extracted with a teaspoon!"[18] Prospecting nearby for additional occurrences of gem-quality tsavorite was carried out by comparing the sizes of termite mounds, the larger mounds indicating softer rocks and easier digging for the termites.[19] Parts of the rich "Crumbly Zone" (**Fig. 1.18**) at the John Saul Ruby Mine were mined with bare hands.

▸ In a study emphasizing the geological similarities between Sri Lanka, famous for its gems since antiquity, and Madagascar, another "Isle of Gems," C.B. Dissanayake and Rohana Chandrajith noted a broad correspondence between primary gem deposits and *concentrations of graphite*,[20] the pure carbon that is the stuff of "lead" pencils. And for a short while, the Merelani tanzanite deposit in Tanzania was mined for its graphite (**Fig. 1.20**). An association of graphite with gemstones exists in Kenya too, and in Madagascar and extreme southern India, and also in Kashmir (**Fig. 1.21**).

Carbon-rich fluids with accumulations of foul-smelling molecules (or their precursors), whatever their source, must have been present *where and when* the gems crystallized. The fluids were therefore subjected to the same conditions as the gems. At the same time, or

Figure 1.20 Crystal of zoisite ("sub-gem tanzanite") with graphite. The zoisite is on the right-hand side of the specimen, the graphite on the upper and middle left. Merelani, Arusha Dist., Tanzania.

Figure 1.21 Crystals of red corundum with gemmy portions, coated by graphite, Batti Kunda (Batakundi), Muzaffarabad District, Azad Kashmir, Pakistan. The greatest dimension of the largest piece is 4 cm. Parts of the interiors of many such crystals are bright red with small gemmy sections here and there. Photo: ©L.-D. Bayle. For additional information concerning on this locality, see http://www.fieldgemology.org/blog_byKey.php?id=59&key=Pakistan

Figures 1.22 and **1.23** Two views of a partially translucent fragment of feldspar found loose on the ground between two nearby ruby mines, Taita-Taveta District, Kenya. Scale in cm.

Figure 1.24 Gem-quality diopside crystals with unusually well-formed graphite from the tanzanite workings at Merelani, Arusha District, Tanzania. The graphite crystals are about 7 mm in largest dimension. Photo: John A. Jaszczak. Specimen in the collection of the A. E. Seaman Mineral Museum, Michigan Technological University.

perhaps later, much of the carbon was transformed into graphite.* One study notes that "significant fluid flow" with "infiltration of carbonic fluids gave rise to graphite schist... frequently associated with formation of tanzanite."[21]

In their recent theses, Garnier[22] and Feneyrol[23] attributed the origin of the graphite, and presumably of the odors, to the metamorphism of marine limestones bearing "organic" matter, or organic matter plus unspecified carbon. This reasonable idea does not, however, readily account for the large dimensions of graphite masses found in Tanzania, Madagascar, and Sri Lanka.

Whether the fluids with the organic odors are of deep origin, or produced locally by the metamorphism of limestones, or a mix of both origins, they accumulated in zones of locally low pressure, as fluids always do. And since the gems crystallized in the same places and at the same time, they too must have grown in zones of locally low pressure.

Hydrogen sulfide (H_2S), sulfur (S_8), some methane (CH_4), and traces of nitrogen (N_2) have recently been identified in fluid inclusions within crystals of tanzanite,[24] while ethane (C_2H_6) and perhaps even propane (C_3H_8) and butane (C_4H_{10}) were noted in earlier studies from crystals found tens or hundreds of meters higher in the same deposit.[25] The presence of molecules of ethane, propane, and butane seems surprising, however, because the temperatures undergone by the tanzanite would presumably have been sufficient to cause them to break down. Yet molecules of skatole (C_9H_9N) have been reported from rocks in the gem fields of Burma.[26]

▸ *Deposits with transparent colored gemstones are not randomly distributed.*[27] Many productive deposits occur in clusters or groups, or in rough alignments more or less parallel to coastlines, or former coastlines, and relatively near the coasts of the southern lands of Brazil, West Africa (Nigeria, Namibia), East Africa (Kenya, Tanzania, Mozambique), and possibly Madagascar, far southern India, and Sri Lanka. Other CCG deposits follow the great arc of the Himalayas, and still others are present along a sector of the Central Urals, locales that were continental coastlines in times past.

▸ There is a qualitative rule that seems to apply to all or many primary gem deposits: the *minerals in nearby rocks are also somewhat better crystallized than is usual.* These include materials such as feldspar, and also mica, talc, and calcite, none of which is suitable for milady's finger. See **Figures 1.22–1.30**. Surface finds of fragments of well-crystallized feldspar and mica were useful as "pathfinders" in prospecting for aquamarine east of Mount Kenya.[28]

▸ The first materials extracted from a primary gemstone deposit are almost always the very best. In Kenya by 1972 or thereabouts, "*the best come out first*" was referred to as Saul's Law. I subsequently learned that Paul Desautels (1920–1991), curator of gems and minerals at the Smithsonian

* Many such reactions involve the odorless gas carbon dioxide.

Figure 1.25 Partly transparent massive talc, Kenya or Tanzania.

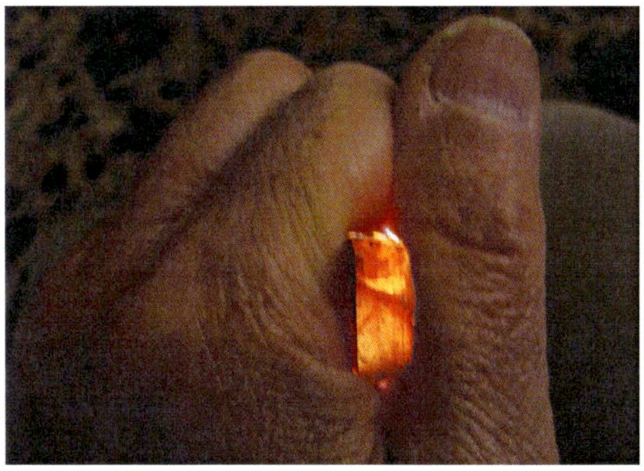

Figures 1.26 and **1.27** Two views of well-crystallized dark mica from near the Golconda mines, Minas Gerais, Brazil. On the right, the specimen is oriented on its edge and is illuminated from below. The Golconda pegmatite produced fine gem tourmaline and beryl.

Institution, referred to the phenomenon as Desautels' Law and that Fred Pough (1906–2006), former curator of mineralogy at the American Museum of Natural History in New York, called it Pough's Rule.

Numerous explanations have been attempted for this "law," not all of them geological. Inexperienced individuals are likely to spot fine gems and to overlook stones of lower quality; prospectors may sell their best stones immediately in order to finance further digging; and surface concentrations of gems liberated from their host rock by natural forces include a high percentage of unfractured gems. Yet the notion I favor is that the best-quality gems are those that had crystallized at the tops of their deposits where the pressure had been lowest, and were thus the first to be discovered. In Mogok, Burma, for example, the best-quality peridot crystals, "known by the Chinese gem miners as 'kings,'" are found on the tops of the primary deposits.[30] See **Figure 1.31**. Gem crystals are commonly associated with indications of intense fluid circulation, especially so at the very tops of the deposits, where the pressure is lowest. Many such fluids will partially dissolve the surrounding mother-rock, producing short-lived fluid-filled cavities particularly suitable for high-quality crystallization.[31]

In short, the tops of gem deposits are associated with relatively low constraining pressure and with enhanced access to fluids with diverse chemical and physical properties. The odors and the massive accumulations of graphite indicate that some of the fluids must have been carbon-based, and inclusions of hydrogen sulfide (H_2S) and of pyrite (FeS_2) in many transparent gems is evidence for the presence of sulfur.

Figure 1.28 Ground covered by well-crystallized flakes of mica (muscovite), a short distance from a gem tourmaline mine, Minas Gerais, Brazil. The presence of the mica would have alerted *garimpeiros* (self-employed prospectors and miners[29]) to the possible presence of a primary gem deposit nearby.

Figure 1.29 Ruby in well-crystallized calcite, Mogok, Burma, ©2008 Ted Themelis.

Figure 1.30 Corundum (sapphire) on a matrix of almandine garnet, Umba River, NE Tanzania. Though neither the sapphire nor the garnet on this specimen is of gem quality, they are both nicely crystallized "sub-gems." The Umba River deposit has produced large quantities of gem sapphires and gem garnets, but, as far as known, no specimens with a gem sapphire and gem-quality almandine together. Largest dimension approximately 2.5 cm. Collection J. Saul.
Photo: ©L.-D. Bayle.

Figure 1.31 "King Peridot" - olivine (peridot), 590 carat crystal from Pyaung Gaung, Mogok area, Burma. "Olivine" is the name of this mineral, a silicate of magnesium and iron, "peridot" is the name for magnesium-rich olivine when it occurs as a gem, and "Pyaung Gaung" means "green color" in Burmese.
Photo: Albert Russ, courtesy of Federico Bärlocher.

Figure 1.32 The geological record begins about 3800 million years ago with the end of the Late Heavy Bombardment ("LHB"), the unique episode, dated 4100–3800 million years ago, that scarred the surface of the Moon and necessarily also the Earth. With exceptions discussed later, transparent colored gemstones (CCGs) were not formed before 660 million years ago. There were, however, several gem-forming episodes thereafter.

▸ With a few interesting exceptions discussed later, there are no primary deposits of crystalline colored gemstones older than 660 million years (Ma*), which is some 80 to 85 percent of the way through our planet's history. In concluding a detailed multiauthored study published in 2013, two of the world's most competent specialists on gemological subjects remarked: "we could ask why so many gems were formed during these times [660–542 Ma]. The answers are not at all evident."[32]

Transparent colored gemstones crystallize within small geographical areas (and not elsewhere), and they did so during distinct geological episodes (and not before and not after). See **Figure 1.32.**

Different types of "gems": A matter of definitions

Deposits of colored gemstones are rare. But as just set out, they are also peculiar in a number of respects, and these *peculiarities are constraints* that must be taken into account by anyone trying to put together a "theory of gemstone formation." No general theory of gemstone formation can be correct unless it takes all these peculiar constraints into account.

As used here, the term *CCGs*, for crystalline colored gemstones, refers to the rare transparent varieties of hard, durable minerals that are normally found as unremarkable "stony" substances.

List 1 (*page 4*) shows selected gem-forming materials by their mineral names. As gems, many of these minerals occur in diverse color varieties, some of which have other names. If crystallized in a transparent manner, for example, the *mineral* corundum (aluminum oxide, Al_2O_3) may produce the *gem* sapphire or ruby, depending on its color. Similarly, if suitably well crystallized, the *mineral* named olivine is called by the *gem* name peridot (**Fig. 1.31**), and the *mineral* beryl may become the *gem* known as aquamarine, but, depending on the color, it may instead merit the name heliodor, golden beryl, morganite, red beryl, or emerald.[33,]

My definition of CCGs is designed to include substances that naturally "belong" together because of the comparable conditions under which they were formed. Other substances, including diamonds and certain emeralds that are treated as "gems" in everyday language by laymen and geologists alike, have been excluded from my definition of CCGs. Aside from minor overlaps, these materials are not found in the same geographical areas as the gem-forming minerals I have listed, nor were they formed in the same geological times.

- Diamonds are excluded from my definition of CCGs because they crystallized at depths of approximately 150 to 250 kilometers, far deeper than the gem varieties of minerals on my lists.[34]
- Emeralds and alexandrites from deposits older than approximately 660 Ma are excluded as special cases, apparently formed in contact with yielding masses of mica and talc that cushioned their growth, thus facilitating their crystallization.
- Emeralds from Colombia formed under unique well-studied circumstances and are not treated here.[35]
- Peridots (gem olivines) that formed in association with volcanic eruptions (most famously from Arizona's San Carlos Apache Indian Reservation) occur as irregular gemmy masses lacking traces of crystal faces.[36] This sets them apart from the other materials listed here, including peridots of other origins, and they too are excluded from my definition of CCGs.
- Jade, turquoise, opal, and agate are not crystalline colored gemstones by my definition because they are either noncrystalline or nontransparent or both.
- Ivory, pearl, and amber are not minerals, nor is lapis lazuli (a mixture of different minerals, i.e., a rock).

* In English, the international scientific abbreviation for million years, "Ma," is read as "million years" or as "million years ago," depending on the context.

- Obsidian, moldavite, Libyan Desert Glass, and other natural glasses are not crystalline.
- Soft minerals such as calcite, gypsum, sulfur, and rock salt commonly form transparent crystals at low, easily attained pressures and temperatures. Such materials are not rare and since both durability and rarity are qualities conventionally attributed to gems, they too have been excluded.
- Rock crystal, amethyst, citrine, and other gems of the quartz family[37] are excluded as a matter of convenience because some deposits formed at temperatures and pressures very much lower than other gem-forming minerals. Transparent crystals of quartz are not rare and are generally low priced.[38]

CCG-forming minerals are hard; there is little value to a ring with a stone that will be damaged by household dust or scratched by fingernails. All the minerals in **Lists 1** and **2** have a hardness of 6 or more on the Mohs Scale of Hardness, which ranges from talc to diamond, and in which a variety of feldspar (orthoclase) is used to define hardness 6, quartz as 7, topaz 8, and corundum 9.

List 2: Crystalline colored gemstones (CCGs)

Revised reference list of minerals of hardness 6 or more that crystallize in a transparent "gemmy" manner under certain circumstances

andalusite	grandidierite	rhodizite
axinite	hambergite	rutile
beryl*	hibonite	scapolite
beryllonite	humite	serendibite
bromellite	hypersthene	sillimanite
chondrodite	jeremejevite	sinhalite
chrysoberyl*	johachidolite	spinel
clinohumite	kornerupine	spodumene
cordierite	kyanite	taaffeite
corundum	londonite	topaz
danburite	magnesioaxinite	tremolite
diopside	manganotantalite	minerals of the tourmaline supergroup
epidote	olivine*	vesuvianite
euclase	painite	zircon
several types of feldspar	petalite	zoisite
minerals of the garnet group	pezzottaite	
	phenacite	

* Emerald (a variety of the mineral beryl), and alexandrite (a variety of chrysoberyl), from deposits older than 660 Ma have been excluded from the definition of crystalline colored gemstones. Those peridots (the gem variety of the mineral olivine) that formed in association with volcanism and which lack crystal faces are also excluded.

List 2, constructed with these constraints in mind, provides a more coherent and natural definition for crystalline colored gemstones than was provided by **List 1**. **List 2** remains somewhat arbitrary, however, for the number of entries could have been increased or decreased by changing the hardness cutoff. Lowering it, for example, would allow the inclusion of transparent varieties of sphene (hardness 5 to 5½), brazilianite (5½), and apatite (5), all of which are occasionally faceted and used as gemstones.

Pressure and temperature and the stability of minerals

Ice, an entirely bona fide mineral with a hardness greater than talc, does not form above 0°C at the atmospheric pressure encountered at sea level. This is in accord with a general rule whereby each mineral normally forms within its particular range of pressures and of temperatures, its "P-T stability field." But every mineral is different and many hard CCG-forming minerals only crystallize at the high pressures that exist at substantial depths within the Earth.[39]

Yet not all of the various CCG-forming minerals crystallize within the same range of depths (pressures), and this raises puzzling questions because *CCGs that had crystallized at different depths are very commonly found together* in gravels and other secondary accumulations produced by erosion of the original (primary) host rocks.

Pressure–Temperature stability fields for many minerals have been established by experiment. It appears, however, that the pressure and temperature ranges obtained in laboratories may not have much relevance when it comes to the formation of CCGs in nature. For as discovered and established by Virginie Garnier and her coworkers, naturally occurring salt (the mineral halite), along with sylvite (potassium chloride), gypsum (hydrous calcium sulfate), and other sulfate minerals, may be present in the impure rocks that characterize the world outside the lab. If so, they can serve as fluxing agents when melted in the course of metamorphism, *permitting other minerals to crystallize at lower temperatures* than those indicated by laboratory determinations made in flux-free and otherwise ideal conditions. In some cases, lower temperatures may in turn imply shallower depths of formation and *lower pressures*.

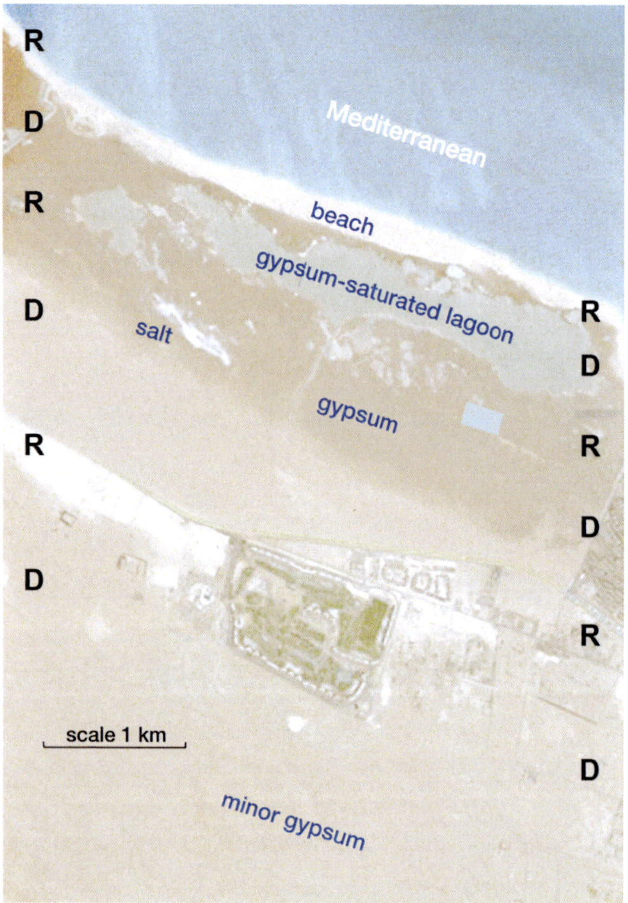

Figure 1.33 A gypsum-saturated lagoon, salt deposits, and deposits of gypsum near El Alamein, Egypt. Ridges (R) and depressions (D) are indicated. Adapted from www.southampton.ac.uk/~imw/sabkha.htm

Figure 1.34 Evaporite lens (Eocene), Wilkins Peak, southwestern Wyoming, USA. Courtesy of dynamic-earth.blogspot.com

At the same time, fluxes also increase fluidity, allowing ions that rarely get together to mix with one another.[40] Garnier and her colleagues showed that the salt and other fluxes encountered in their studies had probably originated as sediments that had been deposited in lagoons[41] or as salt flats,[42] such as those found at the very edges of continents (**Fig. 1.33**), and then metamorphosed. In general, such "evaporite minerals," as they are called, are preserved as thin lens-shaped rock formations[43] (**Fig. 1.34**).

In sum, each CCG-forming mineral has its own P-T range in which it can crystallize, but the existence of irregularly distributed natural fluxes, such as those found in rocks that originally formed as beachside sediments at the edges of continents, may render the situation too complicated to allow us to specify the *temperatures* for the formation of gem-quality crystals. Thus the temperature at which Himalayan rubies formed may have been substantially different from, say, that which prevailed during the formation of the rubies of northern Mozambique. And in both cases, the temperatures would have been different from those required for the formation of, say, tanzanite or topaz.

Further complications arise from the fact that pressure and temperature are not independent of each other. The lowest and highest *pressures* at which a particular mineral would be able to crystallize would not be the same if it had formed at 600°C or at 650°C. Simply put, *the presence of fluxes, not all of which have identical effects, makes it extremely difficult to generalize about the actual real-world P-T conditions at which gems form.*

The problem

Primary deposits of CCGs are peculiar in a number of ways. But in one respect, noted earlier, certain secondary accumulations and deposits seem even more puzzling.

Many CCGs have been recovered from river gravels, where, as is the case with gold, their generally high densities cause them to become naturally concentrated. But there are a number of small gravelly valleys and basins, all of which are located near the edges of continents or continental fragments, in which sharp unabraded CCG crystals occur mixed together with abraded or rounded pebbles of other CCGs. Of the approximately fifty different CCG minerals (**List 2**), some 20 to 35 have been found jumbled together in alluvial deposits and gravels in the restricted drainage areas of Ratnapura (the "City of Gems," Sri Lanka), Mogok (Burma's "Valley of Rubies"), Tanzania's Umba River valley (part of which lies on "Penny Lane"), Tunduru (southern Tanzania), and Ilakaka (Madagascar).[44] Each of these localities — none of which is located deep within a continental interior — has yielded a host of CCGs of diverse types and colors with, in at least one case, "an extreme mingling of ruby and sapphire from different origins."[45]

Yet many of *the diverse CCG-forming minerals crystallized under different pressures,* which is to say *at different depths.* How, then, could they wind up in a single valley or catchment basin? Transport by streams, which is surely a large part of the explanation, does not provide the full answer because some gem crystals retain sharp edges, as though they had barely been rolled by flowing water. This in turn could be partly explained if a river were to cut across a group of gem deposits. In any case, primary gem deposits have been discovered nearby in two or three of these areas (Mogok, Umba, and perhaps Ratnapura).

It appears that *primary hard-rock gem deposits had originally been stacked more or less vertically one above the other.*[46] And since the disagreeable organic odors are present in the original hard-rock deposits, the molecules that caused the odors should also have been locally present at various depths. For some deposits, such as the Himalayan rubies studied by Garnier[47] and the diverse tsavorite garnets studied by Feneyrol,[48] it is likely that the disagreeable odors came from the chemical transformation of organic or sulfur-containing molecules already present in the evaporite rocks, perhaps the remains of microbial life that had flourished in the seas at the edges of ancient continents. And at the Merelani tanzanite deposit, the hydrogen sulfide may likewise come from the breakdown of gypsum ($CaSO_4{\cdot}2H_2O$) during metamorphism of evaporites.[49] Yet evaporites (whose presence cannot be excluded) have not so far been implicated at other primary gemstone deposits, in particular at the John Saul Ruby Mine, Penny Lane (rubies), and Umba (sapphires, almandine garnets; **Fig. 1.30**). At these deposits, the organic odor is associated with metamorphosed igneous rocks[50] whose compositions are similar to the mantle material that underlies the Earth's crust.* These chromium-bearing rocks are rich in the mineral serpentine, which gives them their characteristic "snakelike" greenish color.

Whatever the origin or origins of the disagreeable gases, they would have accumulated wherever the pressure had been locally lowest. And there some of them had been trapped, trapped in the places and at the times the gems crystallized.

Pressure

Well-formed transparent crystals grow in conditions of relatively low or lowered pressure within the P-T range of the mineral in question. This is true in nature, and it is true in the laboratory. In nature, enormous transparent crystals of non-CCG minerals such as gypsum or calcite (hardness 2 and 3, respectively) have been found in caves (**Fig. 1.35**), and a facilitating factor when certain emeralds crystallized was the low resisting pressure of the massive mica (**Fig. 1.36**) against which the emerald crystals grew.

Figure 1.35 Gypsum crystals with large transparent areas, Naïca, Chihuahua, Mexico. The scale is indicated by the group of miners in orange suits. The largest crystals are approximately 15 meters in length, said to be by far the largest known crystals of any mineral. This variety of gypsum is known as selenite, an ancient term that refers to the Moon. Gypsum has a hardness of 2. Painting by Hildegarde Sennac.

In concluding his massive *Geology of Gems*, E. Ya. Kievlenko (1923–2000) noted that "free-space crystallization"[51] is required in order for many minerals to form in a transparent manner. Kievlenko's observation was not novel. In his *Essay About The Origine & Virtues Of Gems. Wherein are Propos'd and Historically Illustrated some Conjectures about the Consistence of the Matter of Precious Stones, and the Subjects wherein their chiefest Virtues reside*, published 1672, Robert Boyle noted that "the best crystals grow in cavities… others which grow in restricted spaces are molds."

Yet CCG minerals of metamorphic origin (*page 4*) generally crystallize under pressures that prevail at depths of many kilometers. Short-lived voids and "free space" do apparently exist at boundaries between ductile and brittle conditions at such depths and also in gashes along zones of deep

Figure 1.36 Massive mica (phlogopite) from an emerald deposit west of Mananjary, Madagascar.

* The Earth's mantle is the dense and highly viscous material that separates the brittle crust from the fluid outer core of molten iron; the mantle accounts for about 84 percent of the Earth's volume.

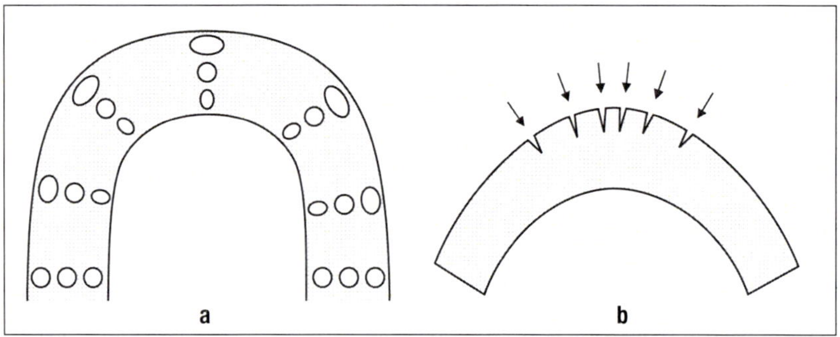

Figures 1.37a and **1.37b** Schematic diagrams illustrating the increase in space and the reduction of pressure on the outsides of folds. Following Sander (1948).

shearing,[52] but the existence of such voids does little to explain the odors, the vertical stacking, or the preferred concentration of CCG deposits closer to coastlines than to continental cores.

Conditions that do not provide actual "free space" may nevertheless produce zones of locally reduced pressure. These include the outer layers of folded rocks; see **Figures 1.37a, 1.37b**.

Zones of locally reduced pressure are also found in the vicinity of "boudins." Boudins, named after a type of French sausage, form when a rock with ductile-rigid-ductile layering is stretched. The rigid layers fracture to form "sausages," and the ductile layers deform and flow to fill the spaces between individual sausages (**Figure 1.38**).[53]

Figure 1.38 Boudins; vicinity of Mount Rushmore, Black Hills, South Dakota. The scale is indicated by the grass in the lower left. Photo: Marli Miller, University of Oregon.

Boudins are common at the Merelani tanzanite deposit in northern Tanzania, where the host rocks for the gems had originally been deposited as flat-lying sediments in a basin, lagoon, or shoreline setting,[54] but were later tilted, stretched and folded. Sausage-string trails of boudins now mark the hinge zones of repeated folds, producing what miners call "ore shoots." Not all boudins contain gems, but the association between boudins and gems at Merelani is so strong that tanzanite prospecting is commonly reduced to little more than a search for boudins. Some of the very richest boudins have been partly mined by bare hands, the fortunate miners emerging covered with graphite.

Folds, boudins, and related geological structures may be useful in allowing gem-quality crystallization to occur, but whereas gems are rare, folds, boudins, and other such structures are not. Their presence in many primary CCG deposits worldwide suggests that they are simply contributing factors. But how, exactly, might they contribute to the formation of gems? They contribute by enabling crystals to grow at pressures that approach the extreme low-pressure boundary of the pressure–temperature range for the particular mineral in question, to grow at the lowest allowable pressure.

This minimal pressure differs for each mineral, and putting a meaningful numerical value on the "lowest possible pressure" is rarely, if ever, possible. First, as we have seen, there is the matter of natural fluxing agents whose effects may vary beyond laboratory experience. Further, there is the matter of temperature itself, for higher temperatures enable certain minerals to crystallize at lower pressures. The "lowest possible pressure" at which crystallization can occur is not a fixed number; it varies according to the temperature. So fluxes, enhanced circulation of fluids, folds and boudins, locally lowered pressures, *and unusually high temperatures* may all contribute to the formation of transparent gems at depth.*

Little more need to be said about most of these factors. But how might unusually high temperatures be attained at depth within the Earth?

Temperature

The notoriously hot working conditions in the deep gold mines of South Africa are not due to special conditions specific to South Africa, but to the fact that the temperature

* Many gems crystallized at very shallow depths in coarse-grained rocks called pegmatites. In certain pegmatite deposits, where extremely low pressures are involved, temperature is apparently not a determining factor in the formation of gems. Gem-bearing pegmatites are discussed later.

increases as one goes deeper in the Earth. Worldwide, this "geothermal gradient," is approximately equal to 25°C to 30°C per kilometer* except in volcanic and rift-valley areas and near the edges of active tectonic plates.

If the geothermal gradient were somehow higher than normal, higher temperatures would be attained at shallower levels, that is to say, at lower pressures. And lower pressures would favor gem-quality crystallization. But as we have seen, there are places where CCGs have crystallized throughout a range of depths. So any additional heat would also have been present throughout the range of depths. A local geothermal gradient substantially higher than normal apparently favors gem-quality crystallization.

A recent (2011) review noted that there is "widespread evidence that ultra-high temperatures of 900°C to 1000°C [1650°F to 1830°F] have been generated" in the Earth's middle to lower crust "repeatedly in time and space." These temperatures were associated with thickened crust in mountain belts that were formed by collisions between tectonic plates, but identification of "the heat source responsible is controversial."[55]

In seeking to identify this "controversial" source of additional heat, we can dismiss familiar factors such as volcanism or rift valleys. Such geological features are far more numerous than gem-producing districts; they are generally not situated in the right places; they do not seem to have been active at the right times in the past; and they are apparently not sufficiently hot either.[56] Higher-than-normal radiogenic heating and "hot-spot plumes" can also be excluded. The required sources of extra heat

- must have affected the entire range of depths of the stacked CCG deposits,
- must have been very localized geographically, and
- must have existed during the limited periods when the CCGs crystallized but not before, and not after.

Here I propose that an additional source of heat comes from friction and shearing generated during oblique collisions between tectonic plates, in particular, from the highly localized grinding of continental plates, one rotating against another. As stated, this proposal is not fully satisfactory, because the spots where frictional heat had been greatest would have also been the places where the pressure had been most elevated, a condition inimical to gem formation. So I am not proposing that CCGs crystallized precisely where the frictional heat had been greatest. But whereas heat circulates and dissipates along irregular paths of lesser resistance, just like a fluid, or is actually carried by a fluid, the same is not true of pressure, which decreases in time and space in the manner of a rapidly fading halo. In consequence, there would be places in the vicinity of points of collisional contact where the temperature had still been very close to the local maximum, but the pressure substantially lowered. It is in such locales, I believe, that gem-quality crystallization occurs.

The unique "Pan-African Event"

In 2003 the journal *Geology* carried an article whose title meant almost nothing to me: "Pan-African is Pan-Gondwanaland: Oblique convergence drives rotation during 650–500 Ma [million-year-old] assembly."[57] The article merited a serious look because my work at the time was focused on what is known as the "Pan-African Event" and the article's title, whatever it meant, suggested that its author, J.J. Veevers, a respected Australian professor, had something new to say.

As students in geology, we were taught that "the present is key to the past." But the Pan-African Event was special. It seemed to be, and indeed was, a unique one-of-a-kind series of heat pulses, episodes that affected substantial parts of the Earth's crust with no counterpart at any time before or since.[58] (Two of the pulses correspond to mountain-building episodes known as the East African Orogeny[59] and the Kuunga Orogeny,[60] the term *orogeny* coming from the Greek *oros* for "mountain" plus *-gen* "to be produced.")

* The temperature increases 5°F for an increase in depth approximately equal to the length of a football field.

"Orogeny" refers to forces and events leading to a large structural deformation of the Earth's crust and uppermost mantle due to the engagement of tectonic plates. Response to such engagement results in the formation of long tracts of highly deformed rock. These tracts are called orogens or orogenic belts. Orogens develop while a continental plate is crumpled and pushed upward to form mountain ranges, and they involve a broad variety of geological processes collectively called orogenesis.

There is nothing in the more ancient geological past of use in understanding the Pan-African Event, and aside from some aspects of Himalayan geology, there may have been nothing like it subsequently. It stands alone.

Our understanding of the Pan-African, and its three or more pulses of heat approximately 660 to 500 million years ago, is far from satisfactory, but we know that the pulses affected the gem-producing parts of eastern Brazil, Nigeria, Namibia, Mozambique, Tanzania, Kenya, Madagascar, extreme southern India, parts of Orissa in northeastern India, and Sri Lanka, and that its effects extended as far as the green tsavorite garnet occurrences of the Sør Rondane Mountains of Antarctica.[61] Yet, as Veevers insisted, the Pan-African did not affect the continental cores, "the cratons," of North America, Europe, or continental Asia.[62]

In sum, the Pan-African heat pulses affected regions *both when and where gem deposits formed*, and it did not, as best as is known, similarly affect the interiors of continents or other parts of the world.

The bumper-car tectonics of Pan-African times

According to Veevers, an oblique continent-to-continent collision some 650 million years ago had set in motion a system of "counter-rotating cogs" (**Fig. 1.39**), generating a sequence of oblique continent-to-continent collisions.

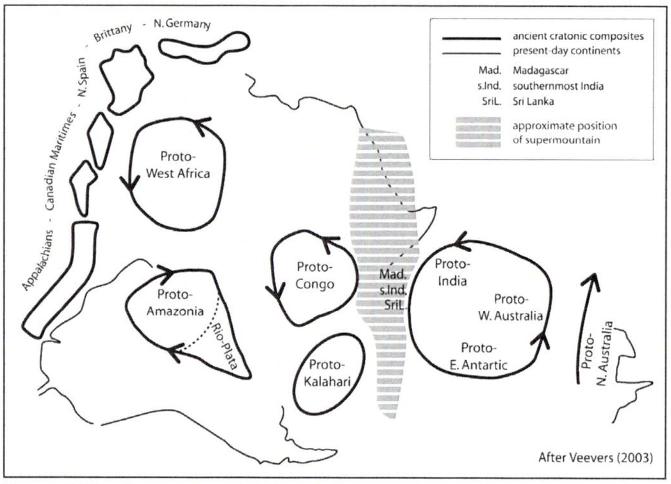

Figure 1.39 Counter-rotating cogs, adapted from Veevers (2003) with additions of the supermountain and the curved and dotted line separating "Proto-Amazonia" from the gemless Río de la Plata.

The continental units, all of which belonged to the western portion of Gondwana,* would in time be known as the Appalachians (which, joined to other land masses, was a sort of baseplate rather than one of the cogs), West Africa, Amazonia, Congo-Tanzania, and Namibia-Kalahari, and the grinding of one unit against another continued intermittently for some 60 to 80 million years. Then around 550 Ma ago, a sequence of less well-understood collisions in East Gondwana began and endured for about the same length of time.[63] East Gondwana then obliquely collided with West Gondwana, generating enormous breaks in the Earth's crust called shear zones, and trapping parts of Mozambique, Madagascar, southernmost India, Sri Lanka, and parts of coastal East Antarctica.[64]

Veevers' proposed cogs (excluding the gemless Rió de la Plata) are indeed roughly cog-shaped, which is to say circular (**Fig. 1.39**). Yet, rather than insisting on their near-circularity here, a task for a later chapter, I wish to emphasize the roughness of the contacts between pairs of Veevers' continental cogs. For it is at irregular points of contact that frictionally generated heat would have been greatest.

Veevers set out evidence for three closely spaced but distinguishable heat pulses dated approximately 650–590, 580–550, and 530–515 million years ago, a spread of ages typical of the dates that have been determined for the crystallization of gemstones in Gondwana. These pulses correspond to three distinct but closely spaced continent-to-continent, cog-to-cog, heat-generating, mountain-building, grinding clashes.[65]

The grand finale involved the amalgamation of the two halves of Gondwana[66] as the western edge of East Gondwana was partly "subducted" (**Fig. 1.40**), passing beneath the eastern edge of West Gondwana. This collision generated frictional heat across great inclined sheets of rock, and was especially intense near points of rough contacts. Ultra-high pressures (discussed later) would have been produced in some places, but elsewhere heat

* *Gondwana,* with *Gondwanan* as the adjective, has replaced the older term *Gondwanaland.* "The Pan-African Event" is so named because its effects were initially recognized in African rocks. Veevers demonstrated that the Pan-African "event" was actually a Pan-Gondwanan series of episodes.

generated by friction and shearing would have been the predominant phenomenon.

The amalgamation of the two halves of Gondwana also produced a north–south supermountain[67] (**Figs. 1.39, 1.41, 1.42**), now eroded down to roots,[68] and these roots host CCG deposits from Somalia[69] in the north to the tsavorite garnet occurrences of Antarctica's Sør Rondane Mountains.[70] Near the suture zone in Tanzania, root rocks were heated to 700°C to 800°C — and even higher locally — while similar rocks located inland and to the west only attained 200°C to 300°C.[71]

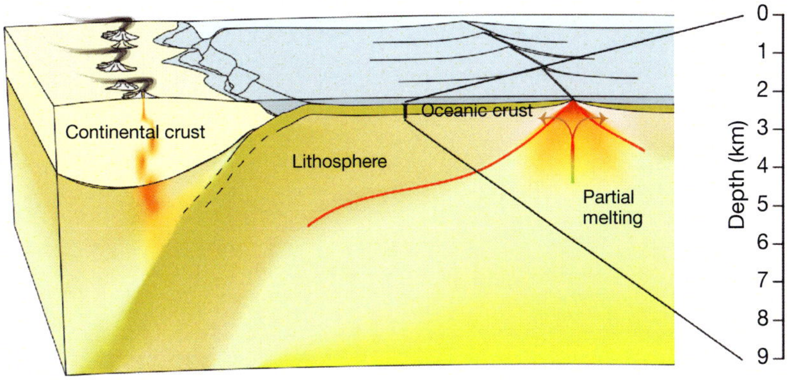

Figure 1.40 Subduction (generalized illustration). Adapted from a figure courtesy of Christopher Smith-Duque.

Concluding his study, Veevers noted that the Gondwanan cycle ended around 500 million years ago with uplift and cooling of "heat emitted during convergence"[72] of the two halves of Gondwana.

Although detailed understanding of the events of the Pan-African remains unsatisfactory, we now know, as noted earlier, that heat pulses affected areas both *where and when* Gondwana's multiple deposits of crystalline colored gemstones were formed.

The alluvial accumulations of diverse CCGs at Umba (northern Tanzania), Tunduru (southern Tanzania), Ratnapura (Sri Lanka), and Ilakaka (southern Madagascar) are derived from rocks heated during Pan-African times. At Umba and Mogok (a post-Pan-African case discussed later), primary CCG deposits have been discovered in the same valleys as the alluvial accumulations.[73] This suggests that in some cases, primary deposits, many of which have been eroded out of existence, had been formed approximately one above another within slanted 3-D contact zones where one plate passed beneath another.

With a few exceptions discussed later, deposits of transparent gemstones are unknown from the first 80 to 85 percent of the Earth's history, from times before the Pan-African. But CCG deposits in the Himalayas, California, the Urals, and Transbaikalia all testify to the occurrence of occasional oblique continent-to-continent collisions during the most recent 15 percent of our planet's history. This suggests that *oblique collisions* between continental plates, *combined with deep subduction*, had not occurred before the Pan-African[74] and that the formation of the Gondwana supercontinent constituted a fundamental change in the evolution of our planet.

With the formation of Gondwana, the workings of plate tectonics appear to have produced new sources of heat and localized heat flow. New phenomena that seem to have emerged in these times included plate-to-plate collisions at oblique angles, the deep subduction of one plate beneath another, introduction into the Earth's mantle of large quantities of intergranular seawater, and local stretching and loosening of the Earth's crust. Earlier continental collisions would have been more frontal, less deep, and perhaps relatively gentle, with an action more readily compared to a floating log bumping against a dock and

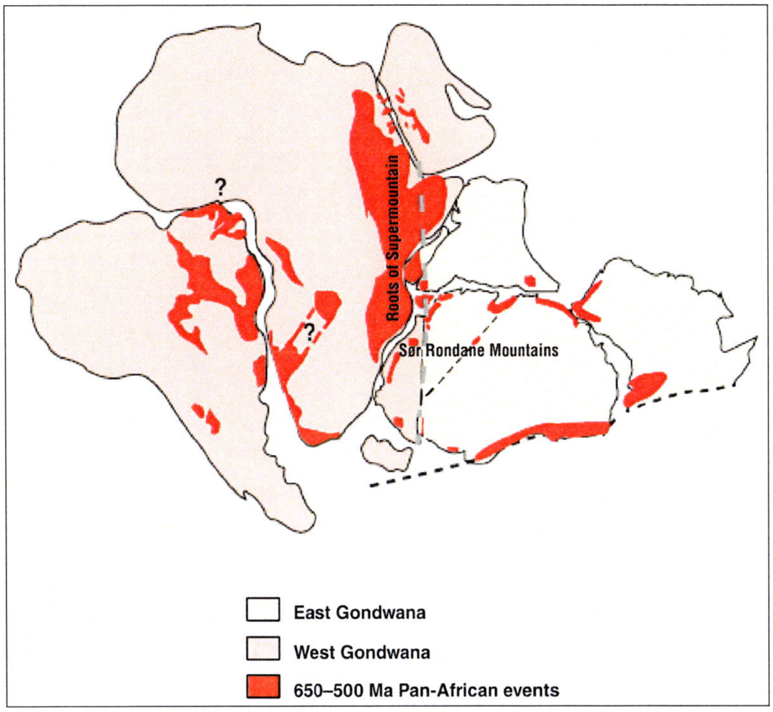

Figure 1.41 Positions of the southern continents about 500 million years ago. At the time, long before the genesis of gems in the Himalayas, Urals, or California, virtually all of the world's deposits of crystalline colored gemstones would have been located within areas marked in red. The Sør Rondane Mountains are indicated by the red area just north of the letter ø. Modified from Grunow et al. (1996).

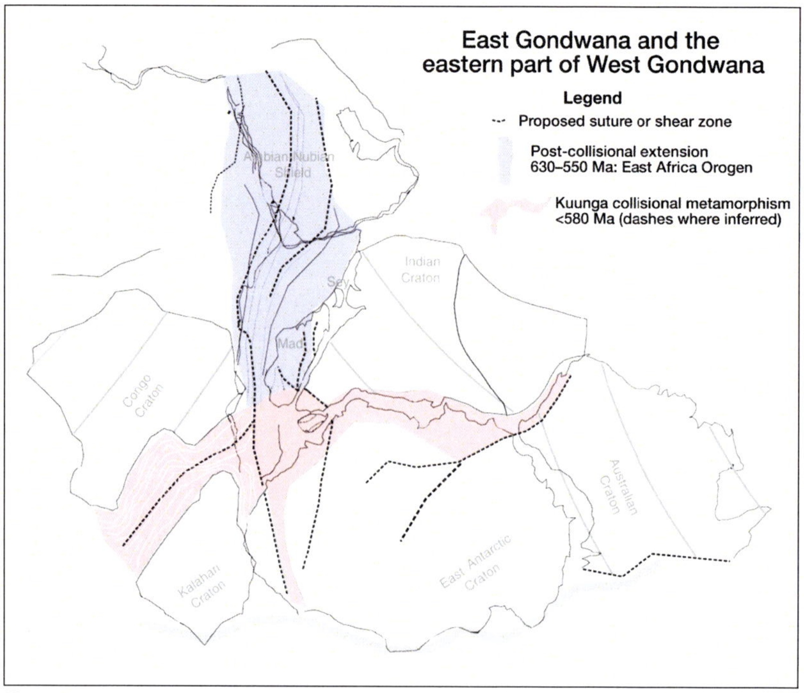

Figure 1.42 Meert (2003) emphasizes two events, an east–west collision associated with the East African Orogeny (in blue), followed by a north–south "Kuunga" collision and orogeny (pink). His reconstruction of events seems relevant to the formation of gem deposits of Namibia, southernmost Madagascar, southernmost India, Sri Lanka, Orissa, and elsewhere, but the Kuunga root zone was relatively narrow and there was no east–west Kuunga supermountain. Modified from a map received courtesy of Joseph G. Meert.

dipping partially under it, than to a deep and forceful grinding of one unit passing below another.

Ultra-high pressure

Events and phenomena discussed above relate to the numerous changes that occur at depth when rocks recrystallize and deform while still essentially solid, that is to say, when they metamorphose. One particular style of metamorphism, recognized only in 1984, is called ultra-high pressure (UHP) metamorphism, briefly mentioned earlier. UHP conditions have been found in rocks that have undergone *deep* subduction and, in common with deposits of crystalline colored gemstones, rocks that underwent UHP metamorphism are absent from the geological record before about 620 million years ago,[79] that is to say, before Pan-African times.

Robert J. Stern has proposed that the absence of UHP conditions indicates that *deep* subduction did not occur in times before the Pan-African.[80] Alternatives may have been subduction in which the leading edge of the subducting plate ceased to descend because of insufficient density, or broke off because of insufficient strength, before attaining great depth.

Loosening of the Earth's crust?

The early Earth was substantially hotter than it is today, and oceanic crust had been relatively thick.[75] In consequence, it may have taken longer then than it does these days for oceanic plates to cool to the point where they were sufficiently dense to subduct.[76] In later times, the cooling of thinner crust (along with its underlying rigid mantle material), combined with deep subduction, would have amounted to a redistribution of the Earth's mass. This redistribution may have been a prelude to a geological scenario proposed by Joseph Kirschvink and coworkers,[77] who contended that exceptionally large-scale reorganization of continental masses occurred around 600–500 Ma ago, and involved rapid continental motions through distances as great as 90°(!). The driving force for these enormous changes, they argue, was a shift in the Earth's spin axis (i.e., a change in the positions of the poles and the Equator). Here (and in more detail in Chapter 3), I further suggest that the shift in the Earth's axis had been caused by the initiation of deep subduction around 650 Ma, after which the Earth's crust became "loose," permitting tectonic plates to travel more rapidly than they did previously.[78]

Pressure, temperature, and the role of fluxes: Conclusions to this point

With the additional heat associated with deep subduction, the lowest permissible temperature for the formation of certain minerals is reached at shallower depths, hence at lower pressures. Reduced pressure facilitates gem-quality crystallization. This occurs throughout a range of depths. Deep subduction also liberates and facilitates the circulation of various fluids. The formation of CCGs may also require friction and fluxing and facilitating factors.

Temperature and Burma's Valley of Rubies

The best gems, or the richest deposits, or the greatest number of deposits, might be associated with the greatest frictional heat. This in turn might be expected where the grinding of tectonic plates has been especially rough. **Figures 1.43, 1.44,** and **1.45** show that the Mogok district of Burma — which is the world's richest by all three measures — is situated immediately to the east of the great near-vertical Sagaing Fault Zone at precisely where (judging from present-day topography) the greatest friction between the Indian Plate and the adjacent (Sunda) plate occurred and is still occurring.

Here, at the western edge of the Mogok Stone Tract some 20 to 16 million years ago, the northward-moving Indian Plate ground against

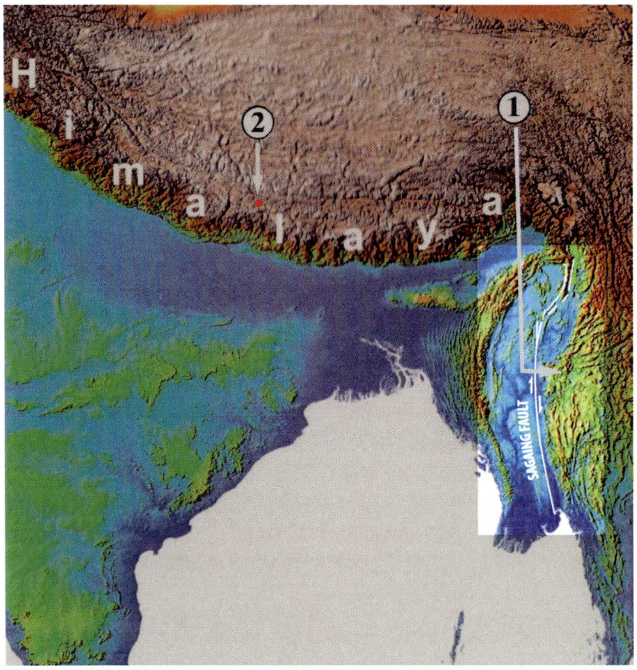

Figure 1.43 Burma, where the world's richest deposits of transparent colored gemstone deposits have been found is highlighted as Area ①. Several occurrences of generally poor quality rubies and perhaps tourmalines have been found in the vicinity of the red mark, indicated ②.

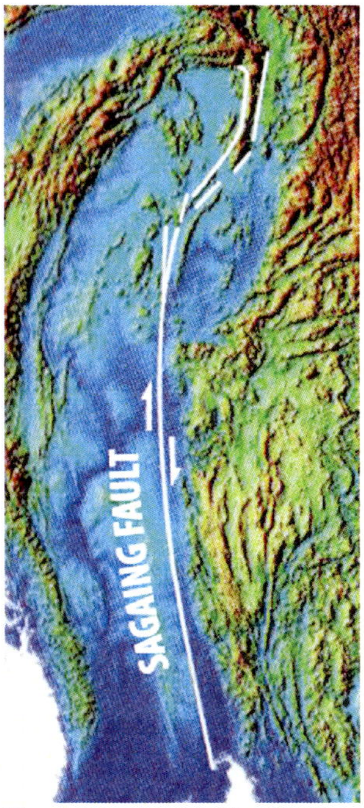

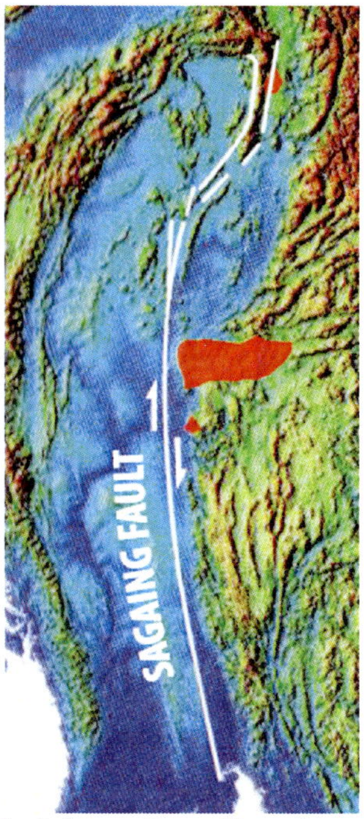

Figures 1.44 and **1.45** The Sagaing Fault, Burma: the positions of major gem-producing areas are shown in red, the largest of which corresponds to the Mogok Stone Tract.

a major peninsula-like topographic asperity, producing "a major thermal event"[81] and a great variety of transparent gemstones, including splendidly transparent rubies whose crystallization occurred about 19 to 17 million years ago.[82]

Although movement along the Sagaing Fault was (and is) mostly vertical and north–south **(Figs. 1.43, 1.44, 1.45)**, the Indian Plate was also being simultaneously subducted downward to the east, a movement that has now been substantially reduced[83] or halted.[84] The subducted slab would have been dragged northward beneath Mogok, causing the two plates to grind together over a large subterranean surface slanted gently downward to the east.[85]

Fragments may have been broken off the Indian Plate and, if so, would have released floods of fluids below Mogok, thereby perhaps accounting for the extraordinarily varied gem and non-gem mineralogy of the area, some 30 kilometers from north to south and nearly 100 kilometers from west to east. Martin Ehrmann (1903–1972), perhaps the greatest "gemological traveler" of the 20th century, remarked that his "own superficial observation" indicated "at least 250 mineral species" in the Mogok area.[86]

The collision of the Indian Plate with the Asian mainland (**Fig. 1.46**) was oblique in the region of Burma (**Fig. 1.43,** Area ①) where the world's richest gemstone deposits are found, but was virtually frontal in the region of the small occurrences of mostly low-grade or sub-gem ruby (**Fig. 1.47**) and tourmaline strung out along the Himalayas in the region of Area ② on **Figure 1.43**.

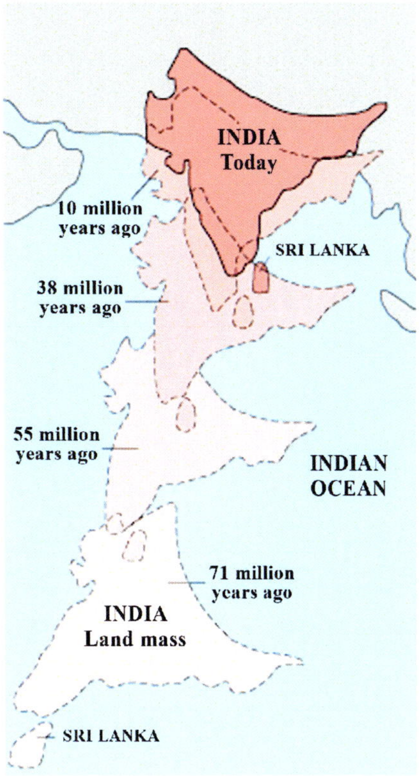

Figure 1.46 The approach and collision of peninsular India with the Asian mainland, with indications of the position of India at various times in the past. Near the extreme western end of the greater Himalayan arc, the rubies at Jegdalek, Afghanistan, crystallized around 25 million years ago. Those at Mogok at the eastern extremity of the collision zone crystallized 19 to 17 Ma ago. These age determinations, carried out on minerals associated with the rubies, are consistent with dates established for the collision of Peninsula India with Mainland Asia. At the Nepalese ruby occurrences of Chumar and Ruyil, situated along the nearly perfectly circular arc of the Himalayas where the continent-to-continent collision has been frontal rather than oblique, crystallization is dated approximately 5.5 Ma and 4.5 Ma, respectively. Dates from Garnier (2003).
Map source: http://www.usgs.org (modified)

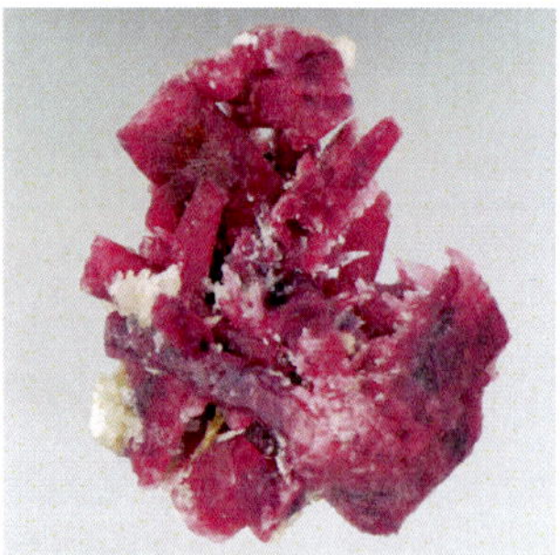

Figure 1.47 Sub-gem ruby, Ganesh Himal ruby deposit, Somdang, Nepal. 6 x 5 x 3 cm. Photo: Mia Dixon/www.PalaMinerals.com

At Mogok, gems were formed through an area from the Sagaing Fault eastward through nearly a hundred kilometers. This perhaps gives an indication of the original widths of gem districts elsewhere.

Pressure and the East African gem deposits

In the jargon of geology, the term *nappe* refers to a large sheet of detached rock that has moved at least one, two, or five kilometers from its original position, depending on whose definition is preferred. Nappes form at depth during shearing when one geological plate collides with another. It was the recognition of stacks of nappes, one above another, that enabled European geologists to make sense of the complex structure of the Alps, now understood to have been formed when the northward-moving African Plate collided with Europe.

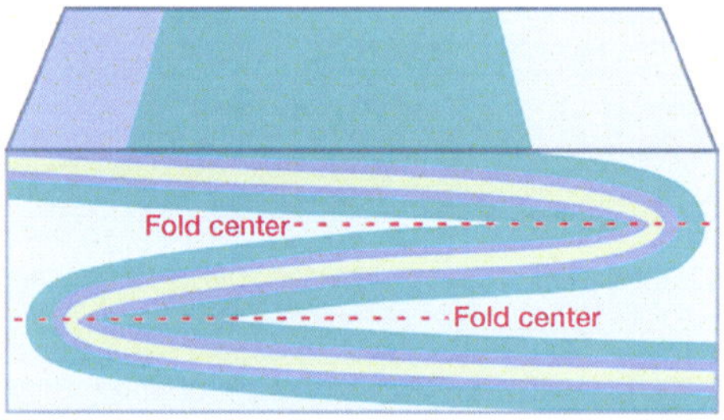

Figure 1.48 Block diagram showing "recumbent" folds. Extreme stress and pressure can cause rocks to shear along a plane of weakness, thereby creating a fault along the fold center. The combination of fault plus displacement results in an "overthrust fault" and if sufficiently displaced, the overthrusted body of rock is known as a nappe.

Figure 1.49 Recumbent folds, Cornwall (Murphy et al. (2008), p. 129).

Figure 1.50 Thrust fault and nappe. Tectonic Arena, Sardona, Switzerland. ©IG Tektonikarena Sardona, Photo: Ruedi Homberger, Arosa.

The original position and geometry of individual nappes can be difficult to decipher, especially in regions, as in East Africa and Sri Lanka, that have undergone subsequent episodes of folding, shearing and erosion, and with the production of "rootless nappes." In all cases, however, nappes start out as tight, more or less horizontal "recumbent" folds that give way and shear, producing great thrust faults (**Figs. 1.48–1.51**).

In my opinion, the oblique collision of East Gondwana with West Gondwana some 550 million years ago is highly likely to have been *the most severe event of any sort* since the Late Heavy Bombardment scarred the Earth and Moon (**Fig. 1.52**) and other members of the Solar System 4100 to 3800 million years ago (with "severity" measured by the amount of energy involved).[87]

The continental crumpling caused by the subduction of part of East Gondwana beneath West Gondwana produced the supermountain, some 8000 kilometers long and more than 1000 kilometers wide, and it did so at a time that Richard Squire and colleagues, whose work is cited here, treat as "a unique window in Earth's history." For in that time, around 700 million years ago, land-based microbes capable of degrading rocks and producing soil had become abundant, but plants with roots that would retain the soil had not yet come into existence. This resulted in extraordinarily rapid weathering of the supermountain.[89]

These days, lower parts of the long-ago supermountain remain as multiple nappes stacked one atop another on the basement rocks of Kenya, Tanzania, northern Mozambique, and Madagascar, overlapping each other in the plains and in the mountains, somewhat in the manner of badly fitted and oddly tilted roof tiles. Rocks of different types and different metamorphic histories are found adjacent to one another.

The actual arrangement and sequence of East African nappes has been disentangled in places, a difficult task accomplished by Yves Deschamps and Elizabeth Le Goff, who observed that each of four highly productive gemstone deposits of northeastern Tanzania — the tanzanite deposit at Merelani, the tsavorite at Lemshuko, the primary deposits of sapphire and garnet at Umba (**Fig. 1.53**), and the Longido ruby mine — was situated high in the uppermost nappe of the local nappe stack. Each of these rich deposits sits at the top of a local sequence,[90] where the local pressure, corresponding to a depth of a few kilometers,[91] would have been lowest.

Similar circumstances pertain at the important Morogoro ruby deposits of east-central Tanzania, the gem deposits at Babati in Tanzania, the fine gem spinel-in-marble deposits at Mahenge, also in Tanzania (**Fig. 1.54**), and at the John Saul Ruby Mine in Tsavo West National Park in southern Kenya.

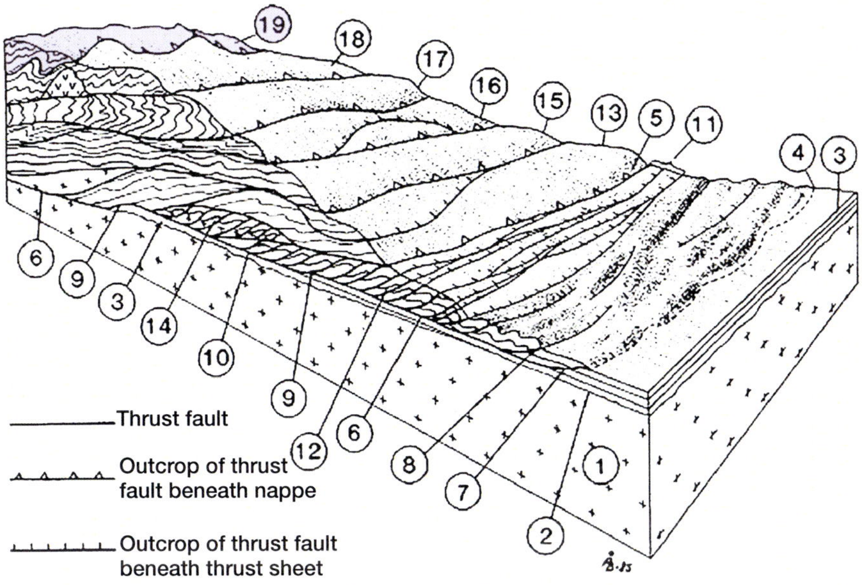

Legend

1. Ancient autochthonous basement
2. Nonconformity surface, peneplain
3. Autochthonous, pinned foreland succession
4. Thrust front
5. Nappe front, present erosion limit for nappes
6. Regional sole thrust
7. Flat ramp, front of sole thrust
8. Splay thrust
9. Ramps, parts of the sole thrust
10. Flat, part of the sole thrust
11. Exposed portion of an imbricate thrust sheet
12. Imbricate nappe developed as a duplex bounded by floor and roof thrusts
13. Nappe with subordinate thrust sheet
14. Duplex with simple small thrust sheets in which basement ① and the oldest component of the foreland succession ③ are preserved

13–19. Nappes which are long-transported relative to underlying tectonostratigraphic units

13, 15, and 16. Nappes having a common history of thrusting

17, 18, and 19. Nappes containing strongly metamorphosed rocks and having an uncertain and in part unknown history of thrusting.

Figure 1.51 Diagram showing possible configurations in which nappes occur and the standard terminology. Older rocks are generally inserted above younger, reversing the normal geological rule of "superposition" in which rocks get progressively older with depth. The history of the uppermost nappes in this schematic diagram is indicated as "in part unknown," suggesting that the geologist who made it wanted to warn that complications are to be expected. Adapted from http://www.ngu.no/upload/Publikasjoner/NGT/pics/fig_27.gif.

Figure 1.52 Surface of the Moon showing impact craters, most of which formed during the Late Heavy Bombardment 4100 to 3800 Ma ago. The lunar surface underwent "impact saturation,"[88] and it is certain that the Earth must have been in a similar state 3800 Ma ago.

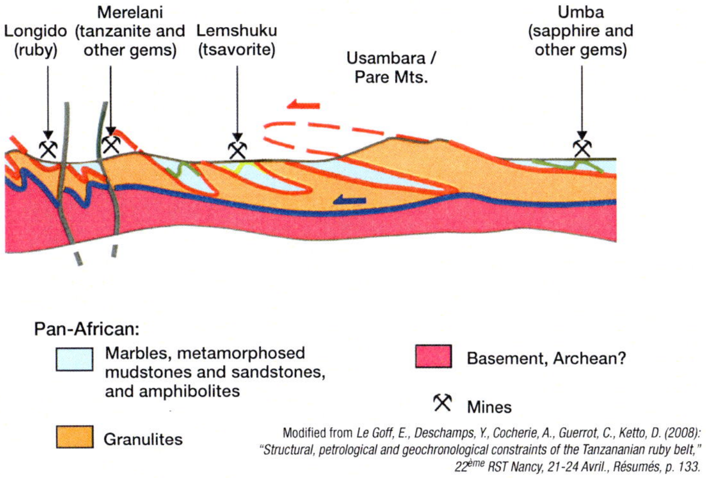

Figure 1.53 Composite geological cross section of the gem-producing area of northern Tanzania (modified from Le Goff et al., 2008).

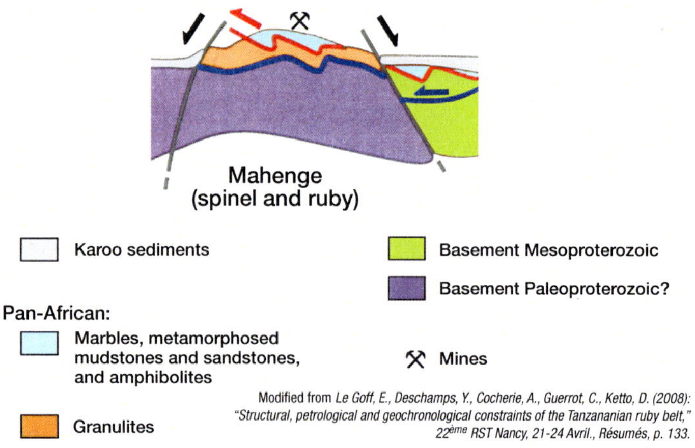

Figure 1.54 Geological cross section of the Mahenge spinel and ruby area, Tanzania (modified from Le Goff et al., 2008).

Hot nappes vs. cold nappes

Tracts affected by continent-to-continent collisions (orogens; *page 14*) do not all host deposits of crystalline colored gemstones. Or if they do, the deposits remain deeply buried. **Figure 1.55** is a reduced black-and-white version of a geological map of Europe, treated so that orogens of various ages are shown as dark areas. Yet significant deposits of gems are only known from a district in the Central Urals near the 2 o'clock position of the roughly circular area framed by these orogens. This suggests that while the presence of orogens and orogenic rocks may be necessary, they are not sufficient to determine the presence of deposits of CCGs. (The circular shapes of large regions partly bounded by orogens is discussed Chapter 3.)

Worldwide, only a few orogens, all of which date from Pan-African times onward, host multiple CCG deposits.[92] Further, those orogens that do host gem deposits may have very extensive gem-free regions. In the East African supermountain orogen, for example, CCG deposits are absent from southernmost Israel into eastern Arabia and Nubia (**Figs. 1.41, 1.42**). Similarly, they are either scarce or absent from central sections of the Himalayan arc, and whereas the Central Urals have long produced a variety of fine CCGs, such occurrences are almost absent from the Polar Urals.

Figure 1.55 Greatly reduced black-and-white version of an originally colored geological map of Europe. The dark areas are for the most part orogens, none of which intrude deeply into the near-circular center of the European continent. Adapted from the Geological World Atlas, 1:10,000,000, Sheet 9. ©C.G.M.W. – Unesco.

Robert Stern has noted that the terminal collision between East and West Gondwana was much more intense in East Africa than in the region of Arabia and Nubia.[93] Harald Fritz and coworkers[94] are more specific, their research highlighting a key difference between the gem-producing areas of the supermountain orogen and the barren zone from southern Israel to Nubia. They show that the gem-rich regions of Tanzania and southern Kenya had been formed in the course of a continent-to-continent collision as such, strictly speaking, while the orogenic area to the north had come into existence through the accretion of arc-like slivers of crustal rocks. The one effect is perhaps comparable to the grinding of one grindstone against another, the other to sandpapering.

- Fritz and his coworkers describe the areas that produce gems as part of a "large and hot orogen"[95] characterized by "thick-skinned thrusting" and the assembly of "a thick pile" of large nappes, perhaps more than a dozen.[96]
- In contrast, the non-gem regions to the north[97] were parts of a "small and cold orogen" with "thin-skinned thrusting," the thinness of such nappes allowing rapid dissipation of heat.

The relevance of hot vs. cold for those interested in the formation of transparent colored gemstones is now evident, though — in common with Veevers and Squire and his coworkers — Fritz and coauthors do not discuss the geology of gem deposits. In passing, however, they characterize as "small and cold" the orogens of the Alps and the Pyrenees, neither of which has produced quantities of CCGs.[98]

Ophiolites

In Kenya, northern Tanzania, Malawi, Pakistan, and perhaps more universally, valuable deposits of CCGs have been discovered closely associated with slices of dark greenish chromium-rich groups of rocks called ophiolites.* In many cases, ophiolites are slivers

* From the Greek *ophis-* plus *-lithos*, for snake-rock, after their usual color.

of oceanic crust that had been caught in the collision of two continental fragments and subsequently trapped within a continental mass.

Ophiolities are particularly rich in minerals of the serpentine family, water-containing silicate minerals with iron and magnesium that, unsurprisingly, are formed during a process termed *serpentinization*. Serpentinization, a multistep process further treated in the next chapter, has other products too, including heat — for the formation of serpentine is heat-productive (exothermic) — and it also generates molecules of hydrogen, H_2.

The hydrogen, a highly reactive substance, does not linger. Instead, it combines with surrounding materials and, among other products, forms the evil-smelling hydrogen sulfide (a.k.a. "stink damp") and methane, the simple hydrocarbon that is the odorless essence of natural gas. If sufficient carbon is locally available, serpentinization may also produce molecules of more complex hydrocarbons — the stuff of petroleum — through purely chemical, nonbiological reactions (Chapter 2).

Slivers of ophiolite are commonly found in the highest levels in nappe stacks. This structural position will regularly cause them to be the first rock units to be removed by physical erosion, and ophiolites are in any case particularly susceptible to tropical weathering and chemical erosion. Indications of their former presence may, however, linger in the bright green of "chrome tourmalines," and the red of rubies, whose color is also due to chromium.

At first the odors that emerge from CCG host rocks seemed inexplicable. It now appears that strong-smelling gases of two different origins — from transformed evaporite minerals and by the process of serpentinization — may have accumulated in places where the pressure was locally low.[99] And in the next chapter we will see that "deep gas" of primordial origin may also be a source of such gases.

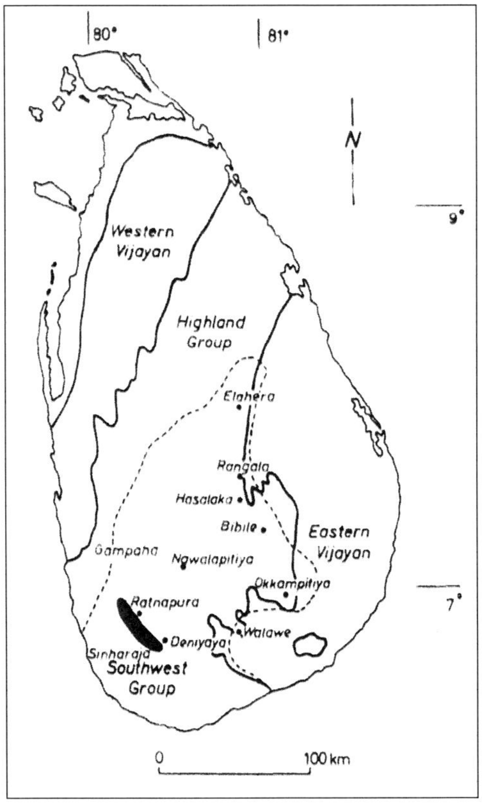

Figure 1.56 Sketch map of the geology of Sri Lanka showing "a probable low-pressure unit" (darkened area) and gem-mining areas (outlined by dashes) as known in 1981. Most of the gemstone mines are in gravels derived from erosion of the Highland Complex (which is composed of the Highland and Southwest Groups). Very few primary deposits are known, all also in the Highland Complex. The great amount of detail on some recent geological maps of Sri Lanka tends to mask fundamental facts incorporated in this 1981 sketch map. Modified from Munasinghe and Dissanayke (1981) fig. 1. The Western Vijayan rocks are now known as the Wanni Complex.

Sri Lanka ("the Isle of Gems") and Brazil

Long before Pan-African times, an elongated basin formed within the ancient "Vijayan" rocks that constitute the 2000 million-year-old basement of Sri Lanka. Erosion from both sides filled the basin with sand and clay[100] that settled and were consolidated, and were eventually metamorphosed during the events of the Pan-African. These rocks are now called the Highland Complex, a curious name for basin-fill, explained by Pan-African faulting and stacking of nappes,[101] and uplift.

Metamorphism of the Highland Complex involved very high temperatures,[102] sufficient to produce a zone of rocks that actually melted (dark area in **Fig. 1.56**). Rocks from this zone were "probably emplaced as a low-pressure" unit,[103] and may belong to an uppermost nappe, for they are also described as situated in "a zone of maximum uplift."[104]

The famous gems of Ratnapura, "the City of Gems," come from gravels derived from the deeply eroded rocks of the melted zone[105] and nearby rocks, as do gems from elsewhere on the island.[106] The Vijayan Group, by contrast, lacks primary CCG deposits.[107]

"Striking similarities" are said to exist between the Highland rocks of Sri Lanka and those of the main gem-producing areas of eastern Brazil.[108]

More on rotating cogs

Collisions involving rotating plates appear able to produce zones of *both* increased and decreased pressure. In an intriguing schematic diagram, modified and greatly simplified here, Leloup and coauthors[109] showed how the collision of the rotating Indo-Chinese plate with South China could produce both compression and extension (with a consequent decrease in pressure). In **Figure 1.57**, the area of extension corresponds to the zone of

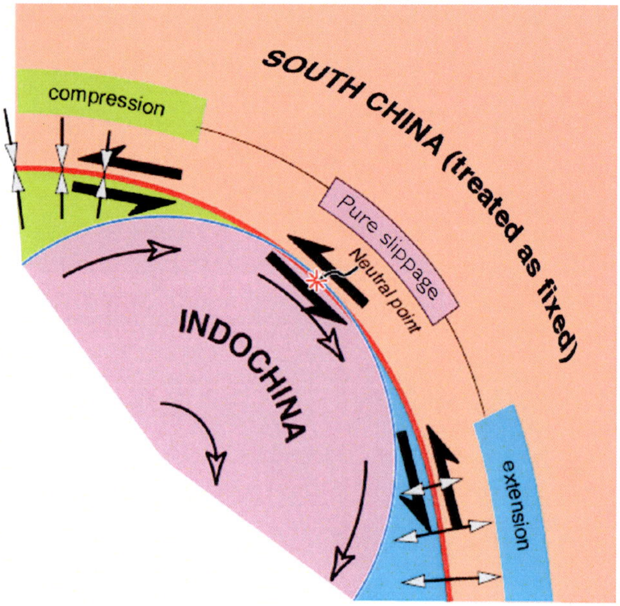

Figure 1.57 Diagram indicating how a dynamic short-lived zone of low pressure (extension) might have formed along the southeastern portions of the gem-rich Red River Valley and fault zone, Vietnam, greatly simplified from Leloup et al. (2001).

faulting and shearing parallel to the Red River Valley, host of numerous deposits of rubies, sapphires, spinels, beryls, and other CCGs.

Recapitulation

The formation of metamorphic deposits of crystalline colored gemstones was favored in times when, and in places where,

- frictional heat generated by continent-to-continent collisions was especially great.
- thrusting, with the formation of large hot nappes, further raised the temperature.
- elevated temperatures throughout a range of depths allowed minerals at different depths to crystallize at lower than usual pressures.
- locally low pressure near the tops of nappe stacks favored the circulation and accumulation of corrosive and mineralizing fluids, as well as carbon-rich fluids with organic odors.
- fluxing agents, derived from sediments originally formed in salty coastal lagoons or lakes, and subsequently mobilized by thrusting and mountain building, were locally available.
- deep faults, tight folds, boudins, rough spots between obliquely colliding plates, or other contributing factors were available.

The need for contributing factors

The finest crystalline colored gemstones in a deposit that formed at a depth of 10 kilometers might occur only within the top 10 meters. In this possibly extreme case, the difference in pressure between the top of the deposit and its lower regions 10 meters below, would be very slight. Considerations of this nature suggest that contributing factors of one type or another may always have to be present.

Crystalline colored gemstones formed in pegmatites

Many deposits of colored gemstones have been found in rocks called pegmatites formed by the upward leakage of concentrated watery fluids derived from granites (**Figs. 1.58–1.60**). Typical pegmatites (or "granitic pegmatites," to be more exact) are smallish slab-shaped, boat-shaped, or podlike bodies that, by definition, have individual crystals larger than 2.5 centimeters (1 inch).

Many gems in areas affected by the events of the Pan-African have been extracted from pegmatites, especially in Brazil, Namibia, and Madagascar, less so in Sri Lanka, where the main deposits are secondary. Pegmatites have also been major sources for CCGs in the geologically younger gem-producing regions of Pakistan-Afghanistan and Burma, California, Maine, the Central Urals, and central Transbaikalia.

Pegmatites are common geological features worldwide, but pegmatites bearing gemstones are rare and appear to be generally limited to areas such as these where a collision of continents or continental fragments is certain to have occurred or can be surmised.

William B. "Skip" Simmons has observed that gem pockets are virtually absent from pegmatites that formed under high-pressure conditions. Generalizing, he went on to conclude that *any phenomenon* that decreases local pressure within a pegmatite would enhance the chances of formation of gem pockets.[110] But in referring to "gem pockets" rather than "gems," Simmons has departed from the storyline I presented in discussing CCGs formed during metamorphism. For despite what I have written so far, pegmatites may produce voids and free space at depth, a matter I treat now, for the most part following the work of Simmons.

The main constituents of granites and of their associated pegmatites are quartz and various types of feldspars and micas. There are certain elements, however, whose atoms do not readily fit into the crystal structures of any of these common minerals. Thus, as a granite solidifies, the remaining melt becomes progressively richer in these "incompatible elements," as they are called, many, but not all, of which are light.

Granitic melts are also initially rich in water, and as solidification proceeds, the melt also becomes progressively richer in water, even to the point of saturation. If saturation does occur, the water is forced to "exsolve." The original melt then becomes two distinct fluids, with the water-rich fluid taking a high proportion of the incompatible elements with it.

Figure 1.58 Typical pegmatite vein in marble, Ambalamahatsara pegmatite field, Madagascar. Photo: Asia Gajowniczek, Spirifer Minerals.

Figure 1.59 Beryl (morganite), tourmaline, lepidolite, and amazonite in a pegmatite vein, Ambalamahatsara pegmatite field, Madagascar. Photo: J. Gajowniczek, Spirifer Minerals.

The light watery fraction will then rise and may form fluid-filled blebs, bubbles, or pockets called miaroles, within which ions can diffuse rapidly. As recognized several decades ago, one feature of pegmatites with miaroles is their emplacement at "*small depths and pressures*," perhaps as shallow as one and a half kilometers.[111]

The shallow depth of emplacement of miaroles is also partly attributable to the presence of the incompatible elements, some of which act as fluxes. Fluxing lowers the viscosity of the watery fluid, and permits it to enter "roof cracks," perhaps rapidly, and with concurrently rapid lowering of the pressure. Prominent fluxing agents include fluorine, boron, phosphorus, and lithium, whose presence sets the stage for the crystallization of topaz (for which fluorine is an essential constituent), tourmaline (which requires boron), and spodumene (lithium), whether of gem quality or not.

Ions of various sorts migrate from the main granitic (silicate) melt into the watery pockets. In consequence, crystals grow inward from the margins of pockets and may develop color bands or other types of zoning, reflecting the continuous changes in their chemical surroundings as various elements are consumed. This produces color-zoning in many gemstones, whether to the delight or to the despair of the lapidary. Changes in the rate of cooling and fluidity may also be important. See **Figures 1.61** and **1.62**.

Figure 1.60 Topaz with transparent quartz and well-crystallized mica (muscovite) from a gem-bearing pegmatite, Sagaing District, Burma.

Some of the finest pegmatite minerals are "floaters" that were never attached to the walls of the pockets. Such crystals formed entirely surrounded by fluids (or gels), and their high-quality crystallization is attributable to the omnipresence of the fluid, for fluids possess little or no internal friction, so when pressure varies in (nonflowing) fluid surroundings, the pressure change is transmitted equally in all directions.[112] See **Figure 1.63.**

Figure 1.61 Tourmaline rough and cut, Madagascar. 5.2 x 2 x 2.8 cm, and 36 x 28 x 10 mm. Photo: Mia Dixon/www.PalaMinerals.com

Figure 1.62 Beryl (aquamarine): zoned crystal, Gilgit District, Pakistan, 2.5 x 2.5 x 4.0 cm. Collection J. Saul. Photo: ©L.-D. Bayle

Eventually, all of the fluxing agents, including much of the water itself, which also acts as a flux, may be consumed by the crystallization of topaz, lithium-bearing tourmaline ("elbaite"), spodumene, and other minerals. This may leave the remaining watery fluid with an extremely high concentration of incompatible elements, including, for example, tantalum, which is 16.7 times as dense as water. (For comparison, lead is 11.34 times denser than water.) Then, as the last of the fluorine, boron, or lithium is incorporated in one mineral or another, the remaining fluid becomes unfluxed. Unfluxed, it is below the temperature at which certain minerals would normally solidify (freeze). It is "undercooled."

Figure 1.63 Spodumene (kunzite) "floater," Nuristan, Afghanistan, 3.5 x 2.5 x 2 cm; 94.45 carats. Photo: Mia Dixon/www.PalaMinerals.com

Solidification, which is to say, freezing, in undercooled liquids can be famously rapid. (The formation of ice in water supercooled to -48°C (-54.4°F) occurs too rapidly to be observed in the laboratory.[113]) In consequence, heavy gem-forming minerals such as manganotantalite ($MnTa_2O_6$) may crystallize near the top of a granite and then have no time to sink despite their high density.[114] As might be expected, manganotantalite (**Fig. 1.64**) and other heavy pegmatite minerals are late to crystallize and are rarely transparent except in small zones.

In most circumstances, the "pockets" in miarolitic pegmatites were never voids. Instead, they originated as blebs or bubbles of watery fluids situated within denser silicate fluids near the tops of still-solidifying shallow granites. (Actual voids may be formed if the fluxed fluid escapes upward from a pocket whose granitic walls had already solidified.)

Yet excepting deposits in Finland and the Ukraine, discussed below, CCG-bearing pegmatite pockets are not known from rocks formed before Pan-African times.

In a highly promising variant view, Matthew Taylor argues that fluxes and undercooling are not necessarily involved. Pegmatites, he claims, may form when colloidal solutions of silica become dense gels. In that case, zoning would be caused by multiple switchings back and forth between gel and solution, a phenomenon familiar from industrial processes. Taylor's scenario would allow for the arrival of diverse materials during intervals of fluid flow. It would also allow for rapid crystal growth, the existence of "floaters," the fracturing and subsequent healing of gem crystals, and multiple episodes of gem formation within a single pocket, all of which are commonly observed.[115]

Figure 1.64 Manganotantalite crystal 1.2 x 1.3 x 1.4 cm, Mozambique. Collection J. Saul. Photo: ©L.-D. Bayle.

In certain pegmatite pockets, gem-bearing or not, heavy crystals and detached fragments of wallrock have been found lying on top of small delicate crystals that show no signs of damage. This argues for formation within a gel. Yet there is a reasonable alternative possibility, for feldspar minerals, which are major constituents of granitic pegmatites, gradually decompose to clays and, as observed in many gem pegmatites, these clays may be flushed out of a pocket by groundwater or other fluids, allowing heavy rocks to gently settle.

Many people, geologists among them, assume that a great amount of time must be required for nature to form durable crystals of gem quality. Yet time does not play any obvious role in the ideas outlined here. In truth, considering the delicate confluence of factors required for the formation of transparent gems, it is easier to argue for the *rapid* formation of gem-quality crystals. Dating the age of crystallization of individual zones within single gem crystals may eventually resolve questions concerning the time required for gems to form.

Gemstone deposits formed long before Pan-African times

Gems that crystallized in times before the Pan-African, even long before, include those emeralds (a variety of the mineral beryl) and alexandrites (chrysoberyl), noted earlier, that crystallized against yielding masses of mica, a very specific and seemingly easily explained circumstance. Such gems were therefore excluded from my definition of CCGs (**List 2**). But does mica really cushion the growth of such gemstones, or is this just a sensible but undemonstrated assumption?

It turns out that many alexandrite crystals, whatever their provenance, are completely covered with phlogopite, a variety of mica. Yet inclusions of mica within such crystals are rare.[116] Commenting, Karl Schmetzer, who has studied alexandrites for decades, observed that "it seems that the alexandrites are growing in an environment where they can move the other minerals — i.e., mostly phlogopite — in front of the growing crystal faces without producing inclusions."[117]

Other gemstones that crystallized before the Pan-African are not so readily written off. Most notable among them are the gem beryl and gem topaz pegmatite deposits at Luumäki in Finland,[118] emplaced around 1650–1700 million years ago[119] in the ancient rocks of what is known as the Scandinavian Shield, and the pegmatite deposits of the same two gems, beryl and topaz, at Volodarsk (Korosten) in the Ukraine,[120] dated approximately 1770 Ma.[121] The overall settings of the two sites are surprisingly similar, for despite their location far from Scandinavia, the Ukrainian deposits are also located in the Scandinavian Shield, in a fragment of the underlying shield uplifted by a mechanism proposed in a later chapter.

In the case of Volodarsk (Korosten), and perhaps Luumäki as well, the gem-bearing zones correspond to what had been giant watery bubbles that solidified within the surrounding granitic magma while it was still locally viscous[122] and presumably yielding.

This is an unusual situation, and in both localities the magma itself had also been unusual, giving rise to "rapakivi granites," a relatively rare type of rock composed of minerals that react to changes in temperature in markedly different manners, some minerals expanding and contracting very much more than others. In consequence, rapakivi granites tend to break apart when exposed to fluctuations in temperature, as indicated by the Finnish word *rapakivi*, which translates as "crumbly rock."

The beryl and topaz deposits at Volodarsk (Korosten) and Luumäki represent another type of special case, in which gems apparently crystallized from the watery fraction of the granitic fluid before the silicate fraction had fully solidified, with the remaining magma itself providing a cushioning effect. But even if solid, this particular granitic material may have had a tendency to break apart, providing "free space" for crystallization.

To recapitulate:

- deposits of crystalline colored gemstones in Gondwana, in the Himalayan region, and elsewhere,
- emerald and alexandrite associated with massive mica, and
- gem beryl and topaz associated with rapakivi granites,

all crystallized under conditions in which the external constraining pressure was substantially lower than normal. Other factors played roles as well.

In addition, there are two CCG occurrences from times before the Pan-African for which I have limited information. The 2810 million-year-old rocks of the Fiskenaesset region of Greenland have produced "a very few facetable rubies"[126] of decent clarity well under one carat, as well as an even smaller number of facet-grade kornerupine gems.[127] And in southern India, blue, yellow, and bicolored sapphires from the Karur District of Tamil Nadu State are reported as "found in" rocks dated 3200–2400 Ma.[128]

Beryl and topaz

Aside from beryl and topaz (and quartz, which I excluded from my working definition of CCGs), no other gem-quality crystals are known from Luumäki and Volodarsk, nor have any other types of CCGs been found nearby. Deposits with beryl and topaz but no other transparent minerals except quartz are also known from Gondwana and Pakistan/Afghanistan, and some gem beryls and topazes are notably larger than other types of CCGs. I once helped a colleague transport a 64-kilogram gem aquamarine that had been mined in Brazil,[123] and one of the world's largest faceted CCGs is a 36,200 carat (7.2 kg) Brazilian topaz; **Figure 1.65**. Large gem topaz crystals have also been found in Burma; **Figure 1.66**. Although the reasons, perhaps relating to fluxing, remain unclear, the suspicion arises that transparent crystals of these two particular gem-forming minerals[124] must be somehow relatively easy to form.[125]

Figure 1.65 Brazilian topaz, 36,200 carats, cut by Elvis Gray and Alan Pobanz in 1990.
Photo by Harold and Erica Van Pelt.

Figure 1.66 Large gem topaz crystal from Mogok, Burma.
Photo by Albert Russ, courtesy of Federico Bärlocher.

Blue-Green-Yellow magmatic sapphires

Much of the world's large demand for moderately priced sapphires is furnished by a variety referred to as BGY, whose blue-green-yellow coloring is due to the incorporation of minor quantities of iron and titanium. More precisely, "BGY," as used here, actually stands for "blue-green-yellow *magmatic* sapphires," formed in molten (magmatic) surroundings;* (**Figs. 1.67, 1.68**).

BGY magmatic sapphires are known from eastern Australia (a prime producer of such stones), the Far East including far eastern Russia (**Fig. 1.69**), Rwanda, Cameroon, Nigeria, northwestern Kenya, Madagascar, France, Germany, China, Cambodia, Laos, Vietnam, Thailand, and elsewhere, some close to Gondwanan gem areas, but mostly not.

BGY magmatic sapphires have been found in a variety of geological settings, all associated with continental volcanism in one way or another: in gravelly accumulations below or sandwiched between lava flows, in ancient and modern-day streambeds and terraces, as surface finds weathered out of lavas and volcanic ash beds, in volcanic breccias and debris flows, and so on. In the volcanically active Lake Turkana region of northwestern Kenya, loose BGY sapphires were collected one by one from the surface of the ground throughout much of the 1980s and 1990s until nomads and herders finally swept the area clean. A very high percentage was of gem or near-gem quality. No deposit, as such, was ever found.

The volcanic rocks with which BGY magmatic sapphires are associated belong to a family of dark, fine-grained continental basalts with relatively little quartz, broadly described as "alkali" (i.e., relatively rich in sodium and potassium but poor in silicon).[129] Yet

* For historical reasons, similarly colored sapphires that crystallized in metamorphic rocks have also been called BGY. Awkward terms such as *BGY-mag* and *BGY-met* may eventually be required to avoid confusion.

despite the systematic association with alkali basalts, several lines of evidence indicate that the ultimate origin of the sapphires must have been in some quite different geological context.

To begin with, sapphire, which is a gem variety of the mineral corundum (aluminum oxide, Al_2O_3)* is not chemically stable when immersed in molten alkali basalt. This is obvious on viewing the corroded and resorbed surfaces on many of the gems (**Fig. 1.67**) and from laboratory experiments as well.[130] In addition, although certainly not globally true, the minute quantities of inert gases trapped in BGY magmatic sapphires from China's Shandong Province have isotopic ratios that indicate a probable initial source in the Earth's mantle, well below the depths at which alkali lavas originate.[131] (The mantle, composed of highly viscous dense material, separates the Earth's crust from the core.)

Figure 1.67 BGY sapphires, mostly sub-gem, Le Coupet, Saint-Eble, Haute-Loire, France. Garnets and zircons are also present in this picture. Many of the stones show resorbed surfaces. The colorless crystal near the center is also a sapphire. Diameter of container: 10 cm. Collection Musée Crozatier, Le-Puy-en-Velay, Auvergne, France, No. 1999.15.22.2. Photo: ©L.-D. Bayle.

It turns out that basalt is not the original host rock for BGY magmatic sapphires and that the sapphires are foreign crystals (xenocrysts) within the lavas. It further appears that all occurrences of BGY magmatic sapphires may be secondary. A few possible exceptions include the primary sapphire deposits in the French Pyrenees studied by Monchoux and coworkers,[132] though it is unclear whether these occurrences are typical of BGY magmatic sapphires. And since we are uncertain what the BGY host rock "should be," it is actually doubly difficult to say whether the deposit in the Pyrenees is a primary occurrence. It seems likely that the materials from which BGY magmatic sapphires crystallized will turn out to straddle a range of chemical compositions, most of which were silicon-poor and more or less rich in aluminum, sodium, and potassium.

Figure 1.68 Zoned BGY sapphire, Nigeria 15 mm x 13 mm x 5 mm. Collection J. Saul. Photo: ©L.-D. Bayle

Eight or more competing scenarios have been proposed to account for the origin of BGY magmatic sapphires and how they wound up in the basalts, a confusing situation, and the confusion increases in comparing the (1) ages radiometrically determined for the solidification of the basalts with (2) ages determined from inclusions within the sapphires themselves. The first thing of note is the spread of ages for the basalts, beginning around 80 million years ago[133] and continuing almost up to present times, producing similar sapphires all along. But some of the sapphires themselves are *much* older than the basalts, having crystallized some 100 to perhaps 300 million years ago[134], thus indicating that some BGY sapphires had resided at depth within the Earth for 200+ million years before being transported to the surface by hot, corrosive basaltic lava.[135] This is puzzling because other BGY sapphires crystallized at more or less the same times as the formation of their associated basalts.[136]

As best as is known, BGY magmatic sapphires crystallized

- from a variety of related host materials whose compositions are only partly known,
- within a broad range of depths in the Earth's upper mantle and crust,

* Gem varieties of the mineral corundum, Al_2O_3, exist in many colors. Rubies are red and all other varieties of gem corundum are called sapphires, frequently with a modifier such as "blue sapphire" or "colorless sapphire."

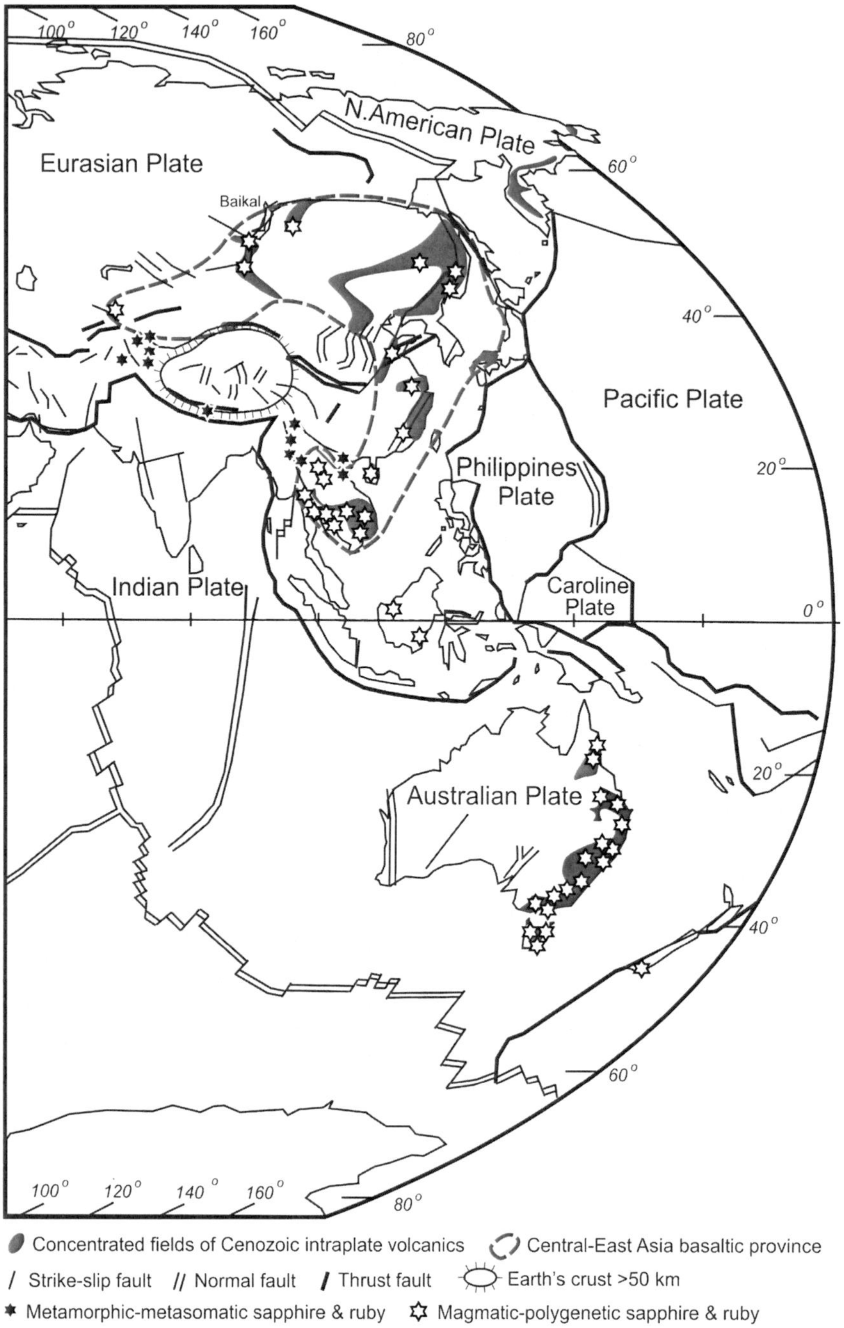

Figure 1.69 Deposits of BGY sapphires in Asia and Australia. Graham et al. (2008); map by Victor Nechaev, Far East Geological Institute, Vladivostok.

- long after the end of the Pan-African, which is to say only within the last 8 percent of our planet's history,
- and were brought to the surface by volcanic materials of various chemical compositions poor in silica (quartz) but rich in sodium and potassium, at various times within the last 2 percent of our planet's history, but not before.

From selected references refined by exchanges with colleagues who have spent years studying BGY sapphires,[137] I initially came to favor a scenario, *which I later abandoned*, in which a type of magma called syenitic (poor in silicon but rich in iron and magnesium) was invaded by a "carbonatite," an unusual sort of geological melt with a composition akin to a limestone.[138]

When mixed with these limestone-like carbonatite melts, such magmas would become oversaturated in aluminum.[139] The resultant inability of the melt to hold aluminum in solution would cause corundum (aluminum oxide) to crystallize, incorporating trace amounts of iron and titanium… or so I reasoned. Then, in another but not necessarily unrelated event, the corundum crystals would be brought to the Earth's surface by a third fluid, the alkali basaltic lava, whether in a massive flow or as a pipe[140] or otherwise.

This scenario seemed reasonable, but the outstanding question, as usual, is why some of the corundum, colored sapphire-blue by iron and titanium, should be transparent, why any gem-quality sapphire should be formed rather than humdrum stony bluish corundum. The answer I originally entertained was that many carbonatites, whatever their ultimate origin, arise from depths that are shallower than most other types of magma. In consequence, I reasoned, the corundum/sapphire could, should, or might have crystallized under relatively low-pressure, hence gemmy, conditions.

This sensible-sounding scenario did not turn out to be satisfactory. For whereas the BGY sapphires in Vietnam and Cambodia do indeed seem to have formed at shallow depths high in the Earth's crust, others, in China's Shandong Province, Laos, Madagascar, Russia, eastern Australia, and elsewhere, have isotopic compositions that indicate origins in the lower crust or upper mantle.[141] My theorizing was further confounded by a study of alkali basalts in central Vietnam by Izokh and colleagues, who, while linking the formation of Vietnamese

BGY sapphires to the presence of free carbon dioxide and to carbonatite-like fluids, were specifically able to exclude a role for carbonatites themselves.[142] This negated my original argument based on the shallow origin of many carbonatite magmas. In that case, the sensible-sounding carbonatite theory could not be right.

Izokh and colleagues also noted that basalts in their study area had entrained fragments of mantle rocks with fluid inclusions composed of virtually pure carbon dioxide.[143] This meant that some or all of the CO_2 involved in the formation of BGY sapphires might not have simply come from a magma in the Earth's crust, but also from the underlying mantle.[144] That in turn would *imply the presence of CO_2 in particular times and places within a very substantial range of depths.*

To recapitulate, BGY magmatic sapphires have been formed during the last 8 percent of the Earth's existence at a variety of depths in the Earth's crust and upper mantle. This occurs in places where magmas that are relatively poor in silicon encounter a source rich in CO_2. But carbonatites, the most obvious geological source for CO_2, seem to have been excluded by the work of Izokh and colleagues, at least in some BGY-producing areas. To this I add a tentative conclusion by Jacques L.R. Touret and Jan Marten Huizenga that "large quantities of mantle-derived CO_2 stored in the lower crust at the final stage of supercontinent amalgamation," may be released into the seas and atmosphere during breakup of the supercontinent.[145]

There appears to be a missing factor of some sort, a geological process or material involving carbon dioxide that somehow allows the mineral corundum to crystallize from certain magmas throughout a broad range of depths, somehow producing gem-quality sapphires. The missing material or factor cannot be particularly rare or exotic because BGY magmatic sapphires are widely distributed (**Fig. 1.69**). Yet it seems not to have been in existence throughout all of geological time, considering that the oldest known BGY magmatic sapphires crystallized within the last 8 percent of our planet's history.[146]

The solution, I believe, lies in another direction, in the existence of deep gas, a matter introduced here, but the central subject of the next chapter.

In a series of scientific articles and a pair of books, *Power from the Earth* (1987) and *The Deep Hot Biosphere* (2001), the Cornell University astrophysicist Thomas Gold (1920–2004) introduced the English-speaking world to a concept current in Russia and the Ukraine for the previous several decades: the Earth is still outgassing great quantities of low-density materials,[147] predominantly methane (the main component of natural gas), whether trapped during our planet's original formation or later during the Late Heavy Bombardment (**Fig. 1.52**). This phenomenon, Gold argued, has consequences that touch on the origins of natural gas, petroleum, coal, limestone, earthquakes, diamonds, water, and life itself.[148]

Might deep gas, carbon dioxide in this case, also be involved in the formation of BGY sapphires? Much of Gold's argument concerns the presence at depth of methane, CH_4, not carbon dioxide, CO_2, but he specifically accepted that local circumstances might favor the existence of deep carbon dioxide rather than methane.[149] Gold's proposal is consistent with observations by Izokh and colleagues of fluid inclusions "of essentially CO_2 composition" in fragments from the Earth's mantle. It is also consistent with the mix of gases trapped in diamonds that crystallized at depths of perhaps 150 to 250 kilometers, which, in decreasing order of abundance in a sample of 28 diamonds from Africa, Brazil, and the United States, were water, hydrogen, carbon dioxide, methane, carbon monoxide, nitrogen, argon, and oxygen.[150]

Gold's proposal could account for the production of BGY sapphires throughout a great range of depths and other geological conditions. The sapphires would crystallize from aluminum-containing (syenitic) magmas that had come in contact with fluid CO_2 and had suddenly become oversaturated with aluminum. Their gem-quality crystallization would be due to the cushioning effect on crystals growing partly or entirely surrounded by a dense fluid. When the pressure varied under such circumstances, the change would be transmitted equally in all directions, with no shearing or folding. Some BGY magmatic sapphires may turn out to have formed in nearly ideal circumstances, as unattached "floaters," entirely surrounded and cushioned by CO_2.

The overall picture would be about the same for all occurrences of BGY magmatic sapphires but the fraction of the corundum that crystallized as gems, the sizes of the crystals, and much else would depend on the chemistry of the aluminum-containing magma, on the presence of other fluids mixed with the CO_2, whether the CO_2 passed through the magma or rose around its edges, and other factors, including the temperature and depth. A second set of variables would be introduced when the sapphires were entrained in the alkali basalts.[151] These would partially or entirely resorb crystals of corundum, gemmy or not, and they might also anneal cracks in some sapphires.

It turns out, however, that Tom Gold may not have the very last word. In the next chapter, in which Gold's hypothesis is examined in more detail, and judged to be to largely valid, the process of serpentinization (*page 21*) is also emphasized as a source for methane and perhaps also carbon dioxide deep in the Earth.

A geologist speculates

The world's most ancient crystalline colored gemstones (CCGs) crystallized approximately 660 to 550 million years ago with the *formation* of the supercontinent of Gondwana. Heat produced by oblique continent-to-continent collisions and the grinding of continental plates, one against another, appears to have been essential for the formation of these gems. The absence of older CCGs suggests that heat in comparable amounts was not generated during the formation of earlier supercontinents.

The world's most ancient blue-green-yellow magmatic sapphires crystallized 400 to 100 million years ago and great amounts of carbon dioxide appear to have been essential for their formation. It is also believed that BGY magmatic sapphires, and the alkali lavas with which they are associated, may owe their existence to the subsequent *breakup* of Gondwana.[152]

Touret and Huizenga proposed that "large quantities of mantle-derived CO_2 stored in the lower crust at the final stage of supercontinent amalgamation," are released into the seas and atmosphere during breakup of the supercontinent.[153] Yet the production of BGY magmatic sapphires has been limited to the most recent breakup. The absence of older BGY magmatic sapphires suggests that CO_2 in comparable amounts was not released during the breakup of earlier supercontinents.

Formation of the crystalline colored gemstones of Gondwana, approximately 660 to 550 million years ago was closely followed by the "Cambrian Explosion" (Chapter 4) when, around 542 million years ago, complex multicelled animals first came into existence. Some 80 percent of our planet's history had come and gone before the first crystalline colored gemstones formed. And the same 80 percent of our planet's history had come and gone before the first complex animals came into being. This cannot be a coincidence.

How much unsuspected information is encased in transparent gemstones?

Notes and references – *Chapter 1: Gems*

1 Partington (1937); Jastrow (1911); and von Dechend (1961, 1966, 1979).
2 Hertha von Dechend, personal communications (1976–1978), for the most part citing Jastrow (1911) and the German writings of Alfred Jeremias (1864–1935).
3 For a variety of examples and illustrations, see Wicke (1984).
4 M.T. Walton (1976).
5 Zodiacs with 4, 8, 12, 13, 16 or 32 sectors have been reported for various peoples and epochs (Saul, 2013).
6 Term defined in Saul (2007a). A more complete definition is proposed later in this chapter.
7 "Cesian beryl" is a mineral name with a modifier indicating that it contains cesium as a trace element.
8 Soils and gravels containing gems may be cemented secondarily by the deposition of limestone, and gems may be encased in lava long after their time of formation. In these and similar occurrences, the encasing rock may be hard but the deposits are of a secondary nature and the term *hard rock* is not used.
9 Bleeck (1908) p. 170.
10 Regional vs. contact metamorphism, cited in Kane and Kammerling (1992) p. 158.
11 Bowerbox and Chamberlin (1995) p. 13.
12 Long recognized; recently investigated by Taylor, Rankin and Treloar (2013).
13 Eckehard Petsch, personal communications (c.1970 and 2007). When queried decades after his visit, Petsch was uncertain which gems had been produced at the mine in question, but thought it had been a tourmaline mine.
14 Gaston Giuliani, personal communication (2005).
15 Iyer (1953) p. 35.
16 C_9H_9N; see Iyer (1953) p. 35.
17 Cédric Simonet, personal communication (2005).
18 Bridges (2007) p. 6.
19 Bridges (2007) p. 5; and C. Bridges, personal communications.
20 Dissanayake and Chandrajith (1999) consider the Mozambique Belt as a belt of both graphite and gemstones.
21 Fritz et al. (2009), citing work carried out by TanzaniteOne Mining Ltd. in 2006.

Specimen of graphite-rich rock from the Merelani tanzanite deposit with bright green crystals of gem-quality grossular garnet (tsavorite), and nearly colorless crystals of gem-quality diopside. Longest dimension 6 cm.

22 Garnier (2003).
23 Feneyrol (2012).
24 Olivier (2006); Giuliani (2010).
25 Malisa et al. (1986); Malisa (1998).
26 See Iyer (1953) p. 35.
27 Saul (1970, 1984, 1985, 2007).

28 Gem-quality crystals of apatite (hardness 5) were also found, some large enough to serve as small stools.

29 The word *garimpeiro* comes from Brazilian Portuguese.

30 Themelis (2008) p. 107.

31 I thank Federico Pezzotta for his useful suggestions and discussions concerning fluid circulation.

32 Rondeau and Fritsch (2012) p. 140.

33 Use of informal or traditional gem names rather than the more rigorously defined mineral names can lead to confusion. The gem-name "tanzanite," which replaced the Swahili *skaiblu* [sky blue], is supposed to refer to the blue-purple gem variety of the mineral zoisite. But yellow, pink, green, and colorless gem-quality zoisites, which are also referred to as tanzanite, have subsequently been found along with the blue and purple varieties.

34 The gemstone trade makes a distinction between "diamonds" and "colored stones." Thus, by convention, a blue or yellow diamond would not be called "colored gemstones," and colorless beryl (goshenite), colorless tourmaline (achroite), and colorless corundum ("colorless sapphire") etc. fall into the domain of "colored stones," an accident of historical usage.

35 For the formation of the emeralds from Colombia's unique deposits, see Cheilletz et al. (1994); and Ottaway et al. (1994). These deposits are not associated with other types of CCGs.

36 Hyrsl (2011) describes such olivine gems as composed of "rounded ultrabasic xenoliths in basalts, typically containing a mixture of olivine, clinopyroxene, orthopyroxene and chromite. Olivine of this genetic type … is known from Arizona, Hawaii, Norway, Czech Republic, Ethiopia, Tanzania, China, etc." The generally larger well-formed crystals of olivine gems from Zabargad (St. John's Island, Egypt), Sapat (Pakistan), and the Mogok area of Burma are treated as CCGs. Gem olivines (peridots) from these localities are found with crystal faces.

37 Prasiolite, ametrine, smoky quartz, morion…

38 Exceptions include fine-colored amethyst and large crystals of colorless transparent quartz suitable for the manufacture of crystal balls and the like.

39 San Carlos type peridots formed near the Earth's surface and were excluded from the definition of CCGs. Many, but certainly not all, quartz gems also formed near the Earth's surface but as a matter of convenience, all quartz gems have been excluded.

40 Garnier et al. (2008). Many gems have been formed in metamorphosed sediments. In some cases, the sediments had been deposited in bodies of water that occasionally dried out, leaving layers of evaporites rich in salts. The presence of salts lowers the temperature at which certain minerals can crystallize. As is known from studies of ruby deposits in the Himalayas and Southeast Asia initiated by Virginie Garnier, the fluxing action of the salt also mobilized certain metallic ions by lowering the viscosity of associated fluids. The presence of the salts also facilitated melting, thereby freeing space for the growth of gem-quality crystals. Ruby is an oxide of aluminium, Al_2O_3, in which the red color is caused by the presence of small quantities of chromium. The mixing or mobilizing of chromium and aluminum is facilitated by the presence of salt and other minerals deposited during evaporation. Also see Suwa et al. (1979); Schwarz et al. (2007); Garnier et al. (2006); and Giuliani et al. (2011).

41 L. Walton (2004) reports that in East Africa, these sediments were originally deposited 2000 to 1800 Ma ago.

42 Feneyrol et al. (2011), who mention sabkhas.

43 Suwa et al. (1979) noted possible evaporite beds on Mgama Ridge, along which numerous occurrences of gem-quality tsavorite and near-gem tourmaline have been found.

44 Ilakaka and "Tunduru" (which includes Tunduru itself and Liwale) are paleoplacers into which dense minerals were washed and concentrated at times when the drainage patterns and topography were different from those of today (Jim Clanin, unpublished).

45 Giuliani et al. (2007) p. 264, who were discussing Ilakaka. Experience suggests that their conclusions can be generalized.

46 Muhongo et al. (2001) write that the "most distinctive feature" of the central domain of the Mozambique Belt orogen "is the abundance of gemstones of great varieties and of good quality." Citing ages between 650 and 550 Ma, they write that "rocks forming under the same broad metamorphic conditions but at different crustal levels (i.e., slightly different pressures and temperatures) display different ages for the peak metamorphic event that generated them."

47 Garnier (2003).
48 Feneyrol (2012).
49 D. Taylor et al. (2013).
50 Mercier et al. (1999).
51 Kievlenko (2003) p. 400.
52 With thanks to Gaston Giuliani for drawing these matters to my attention.
53 For the role of boudins and pinch-and-swell structures at the Sandwana emerald deposits of Zimbabwe, see Zwaan (2006).
54 Giuliani et al. (2013) p. 81.
55 Clark et al. (2011) pp. 235, 239.
56 For the insufficiency of radiogenic heating, see Sawyer et al. (2011) p. 231.
57 Veevers (2003).
58 See, for example, Jackson and Ramsay (1980).
59 Stern (1994).
60 Meert et al. (1995).
61 Osanai et al. (1990); Feneyrol et al. (2013).
62 Terranes in North America, Europe, or Asia with rocks of Pan-African age should be regarded as exotic terranes, transported from sites of origin in Gondwana (Veevers (2003), who cites Camacho et al. (2002)).
63 Veevers (2003).
64 Following Veevers (2003). Meert (2003) has a somewhat different reconstruction.
65 Published dates that vary somewhat from those given by Veevers have been compiled by Feneyrol (2012).
66 Here I follow Veevers (2003). Meert (2003) argues for the sequential docking of individual parts of East Gondwana against an already consolidated West Gondwana.
67 Squire et al. (2006).
68 Known as the "Mozambique Belt." Part of the Mozambique Belt straddling the border of Kenya and Tanzania east of Mt. Kilimanjaro, is known as "Penny Lane" and also as the "East African Gem Belt," but this second term is indiscriminately applied to less well defined areas.
69 http://weerar.wordpress.com/2008/03/16/rich-gemstone-potentials-discovered-in-somaliland/ (6 August 2011).
70 Osanai et al. (1990). It is not clear whether the tsavorite garnets from Sør Rondane, some of which are reportedly 20 cm in diameter, contain portions of gem quality. Commercial mining activity is prohibited in Antarctica.
71 See Fritz et al. (2009) p. 167.
72 Veevers (2003).
73 The enormous alluvial mining zones of Tunduru and Ilakaka reflect more complicated geological histories with multiple episodes of uplift, re-erosion, and re-accumulation.
74 The 2600 million year old emeralds from Zimbabwe are associated with the collision of the Zimbabwe craton with the Transvaal craton (Zwaan, 2006) but the collision does not appear to have been oblique.
75 Moores (2002); Stern (2007).
76 See Stern (2007).
77 Kirschvink et al. (1997).
78 I further speculate that a continental mass that had long been stable had "snapped," and that this imparted the initial spin to one or two continental plates.
79 Jahn et al. (2001).
80 Stern (2005). Arguing along somewhat different lines, Touret and Huizenga (2012), who give multiple references, write: "The first evidence of present-day subduction is marked by the occurrence of Proterozoic eclogites e.g., Lofoten and Madagascar. They indicate the beginning of a plate tectonic system, which closely resembles that of modern time in a probably hotter environment."

81 Bertrand and Rangin (2003) p. 1141. In the Mogok-Mandalay region, rocks of the Mogok Metamorphic Belt are nearly adjacent to the Sagaing fault zone, and are gem-bearing. Farther north and farther south, the Sagaing fault zone and the rocks of the Mogok Metamorphic Belt are separated by tens of kilometers.

82 Garnier (2003); Themelis (2008).

83 Tsutsumi and Sato (2009) indicate vertical motion that is less than a fifth of the horizontal motion.

84 Gahalaut et al. (2013).

85 Le Dain et al. (1984).

86 Martin Ehrmann (c.1960?), cited at http://www.palagems.com/ruby_mines_mogok3.htm

87 Two colleagues reacted in diametrically different manners to my statement. One wrote "Absolutely!" in the margin, while the other commented "How do you know that? … Just because we know less about such [earlier] collisions doesn't mean that they were not as big as the one you mention here." Derry (2006), p.1387, used a related argument, also employed by Squire et al. (2006) in the article in which they coined the term *supermountain*. Derry asked "If continental collision events alone are responsible for imparting high $^{87}Sr/^{86}Sr$ ratios to the oceans, why haven't we seen episodic fluctuations to high values before 600 million years ago, associated with known earlier collisions?"

88 Fassett et al. (2011).

89 Squire et al. (2006).

90 See Deschamps et al. (2006); Le Goff et al. (2006).

91 Deschamps et al. (2006).

92 Some emeralds and alexandrites that were excluded from my definition of CCGs are associated with older orogens.

93 Stern (2008) p. 41.

94 Fritz et al. (2009). Stern (1994) recognized that the terminal collision between East and West Gondwana had been more intense in the south than in the north.

95 Also see Tsunogae and Santosh (2010).

96 Fitzsimons and Hulscher (2005) fig. 5.

97 The relationship of the complex collision of East and West Gondwana to the formation of the emerald deposits of Egypt's Eastern Desert, and to the peridot of the Red Sea island of Zabargad, is unclear.

98 Fritz et al. (2009) p. 171.

99 Zubkov (2009) has written in Russian on "Hypotheses of heavy hydrocarbons and bitumen genesis in different-age ophiolites."

100 See Munasinghe and Dissanayke (1981).

101 Grantham et al. (2008); Kriegsman (1994).

102 http://www.ifs.ac.lk/pdf/lectures/geo/Geology%20of%20Sri%20Lanka.pdf (retrieved 13 February 2013).

103 Munasinghe and Dissanayke (1981) p. 1220.

104 Vitanage (1972) fig. 2.

105 Munasinghe and Dissanayke (1981) p. 1220.

106 Munasinghe and Dissanayke (1981) p. 1220.

107 See Munasinghe and Dissanayke (1981) p. 1217. The Western Vijayan is now called the Wanni Complex.

108 Munasinghe and Dissanayke (1981) p. 1223, and sources they cite.

109 Leloup et al. (2001).

110 Simmons (2007) pp. 186, 190.

111 Ginzburg (1984), who refers to earlier Soviet publications from the early 1950s onward.

112 M. Taylor (2006), whose views have gained much acceptance, argues that the last materials to crystallize may do so within a gel, not a fluid.

113 http://www.labnews.co.uk/news/supercooling-water (retrieved 8 January 2012).

114 See Webber et al. (1999); Simmons and Webber (2008); M. Taylor (2006).

115 M. Taylor (2005, 2006). Granitic pegmatites commonly contain two varieties of quartz, large well-formed gemmy crystals and massive opaque quartz as well. According to Taylor (2006), the transparent crystals of quartz might form from the colloidal solution, with the massive non-gem quartz forming from the gel during final cooling, thereby producing the late-forming quartz cores present in many pegmatites.

116 Karl Schmetzer, personal communication (2 February 2012).

117 Karl Schmetzer, personal communication (2 February 2012); also see Schmetzer and Malsy (2011).

118 Karelia Beryl Mine pegmatite, Kännätsalo, Luumäki, Southern Karelia, Etelä-Suomen Lääni, located 180 km northeast of Helsinki, about 40 km from the Russian border.

119 Lahti and Kinnunen (1993). Peter Lyckberg, personal communication (20 April 2009), refers to age determinations of 1635–1653 Ma obtained more recently.

120 Volodarsk-Volynskii (Volodars'k-Volyns'kyy; Wolodarsk-Wolynskii; Korosten), Zhytomyr Oblast, west of Kiev.

121 Lyckberg et al. (2009) p. 477.

122 Touret (1992); Peter Lyckberg, personal communications (March 2012).

123 This entirely transparent but somewhat fractured boulder appeared to be lovely medium-dark blue, but the intensity of the color was essentially due to its great thickness. A 20 carat stone facetted from a detached fragment barely showed any color at all. Lyckberg et al. (2009) report a flawless 23 kg gem beryl from Volodarsk, 55 cm tall.

124 Lyckberg et al. (2009), p. 496, report that beryl and topaz are "antagonistic" minerals at Volodarsk; where one grows to size, the other dissolves and may be absent or deeply etched. This is not a universal rule, and Lyckberg in a personal communication (January 2012) adds that some of the gem pegmatites in Pakistan, including the well known localites of Shengus and Shigar, did not have the same sequence of mineral formation, and that well-formed crystals of both topaz and beryl may be present on the same specimens at these localities.

125 The mineral spodumene, $LiAl(SiO_3)_2$, also forms very large gem crystals, though smaller than the largest beryls or topazes, but spodumene, large or small, gem or not, only forms if lithium is available.

126 Anette Clausen, oral presentation, October 2013, "Gemstones of Greenland," at the 33rd International Gemmological Conference, Hanoi. Also see her abstract. The Greenland occurrences are noted by Stern et al. (2013, including supplementary material in their Data Repository item 2013202), who also give dates long before the Pan-African for "rubies" from Froland (Norway) and Bahia State (Brazil), but William Heierman, who I thank, and sources in the gemstone trade indicate these localities have only produced non-gem red corundum. For fine photographs and general information concerning gem and non-gem corundum from deposits worldwide, see http://www.corunduminium.com

127 Held by the Geological Museum in Copenhagen and in private collections.

128 Panjikar and Panjikar (2013).

129 Izokh et al. (2010) p. 731; F.L. Sutherland, personal communication (December 2011).

130 Graham et al. (2008); Yui et al. (2006); Green et al. (1978).

131 Hu WenXuan et al. (2007); Graham et al. (2008); Green et al. (1978); I-Ming Chou (2010); Limtrakun et al. (2001); and personal communications from Gaston Giuliani (2011) and F.L. Sutherland (2011).

132 Monchoux et al. (2006); Pin et al. (2006); and F.L. Sutherland, personal communication (October 2011). The sapphires occur in dikes in peridotite.

133 Sutherland (1999) obtained a date of 81.3 ± 4.2 million years from an *in situ* sapphire-bearing volcanic breccia in the New England gem field, New South Wales.

134 Brian Upton, personal communication (18 March 2013) suggests an age of 315 to 300 Ma for the sapphires at Ruddons Point, Scotland. F.L. Sutherland, personal communication (August 2011); also see Zaw et al. (2006). Zircons dated 400 Ma and arguably relatable to sapphire formation have been found at Tumbarumba, New South Wales (Graham et al. 2004).

135 Graham et al. (2008).

136 Garnier et al. (2005); Sutherland et al. (2002) as cited in Izokh et al. (2010); and Graham et al. (2008), but F.L. Sutherland, personal communication (2013) also suggests several possible explanations.

137 Graham et al. (2008); Green et al. (1978); Hu WenXuan et al. (2007); I-Ming Chou (2010); Limtrakun et al. (2001); and personal communications from Gaston Giuliani (2011, 2013), F.L. Sutherland (2011, 2012), and Brian Upton (2013), all three of whom I especially thank.

138 Guo et al. (1996).

139 Guo et al. (1996).

140 Hollis et al. (1983); Robertson and Sutherland (1992).

141 Izokh et al. (2010) p. 719; Robertson and Sutherland (1992); Nechaev et al. (2009); Giuliani et al. (2009).

142 Izokh et al. (2010) p. 731.

143 Izokh et al. (2010) p. 731, citing a large number of earlier studies.

144 Izokh et al. (2010) p. 731.

145 Touret and Huizenga (2011).

146 BGY sapphires have been indirectly dated, with the oldest dated 100, 200, 300, or possibly even 400 Ma, as inferred from the ages of zircons whose own crystallization is thought to be relatable to the crystallization of the sapphires themselves. See, for example, Graham et al. (2008). In many BGY deposits, small numbers of rubies are found along with the more numerous sapphires and zircons.

147 For references on terrestrial outgassing and deep gas including many older entries by Russian scientists, see http://phe.rockefeller.edu/news/archives/384

148 Gold (1999) argued that unicellular life had its origin at depth within the Earth.

149 Gold (1987) ch. 2.

150 Melton et al. (1972).

151 The presence of any remaining CO_2 would substantially lower the viscosity of the basalt.

152 Graham et al. (2008); Veevers (2001).

153 Touret and Huizenga (2011).

2

It is remarkable that in spite of its widespread occurrence, its great economic importance, and the immense amount of fine research devoted to it, there perhaps still remain more uncertainties concerning the origin of petroleum than that of any other commonly occurring natural substance.
—Hollis D. Hedberg, 1964

Deep Carbon and Outgassing

Most types of petroleum contain complex molecules of definite biological origin. Yet the ultimate origins of oil and gas are unclear. For, in detail, the textbook explanation that gas and oil come from decaying marine life and microbes turns out to be unsatisfactory. Sensible arguments can also be made that methane, the principal constituent of natural gas, was trapped in the Earth at the time of its formation ... or that methane and closely related materials were added to our planet in the course of a great bombardment ... or that the Earth regularly manufactures methane and other "organic" molecules at depth or within subducting slabs of sedimentary rocks. The evidence is open to clashing interpretations, in part because of our present-day inability to determine whether samples of the simple molecule methane, CH_4, are, or are not, of biological origin.[1]

Not very long ago, geologists believed the Earth had come into existence as a great ball of molten rock. I heard the Fireball Earth hypothesis as a student in the 1950s and 1960s but do not recall whether I, or anyone around me, actually believed it. But aside from the biblical account, which is not part of the geological curriculum, there were no other theories.

If all or much of the Earth had been molten, any methane then present should have been oxidized (burnt) as it bubbled upward through hot oxygen-containing magma and lava. This would have produced carbon dioxide and water, the natural oxidation products of CH_4. Large amounts of CO_2 and H_2O do in fact spew out of volcanoes. But our planet also possesses great quantities of unoxidized (unburnt, reduced) carbon, one estimate of which gives 200 tons of unburnt carbon for each square meter of the Earth's surface.[2] The existence of these unoxidized materials would negate the Fireball Earth hypothesis unless there was a mechanism to generate unburnt carbon all over again. Photosynthesis is such a mechanism.

When photosynthesizing plants and cyanobacteria die,* they and the microbes and animals that feed on them may settle to the bottom of the sea and be buried. There, subjected to pressure in the absence of oxygen over long periods of time, they are in part transformed into gas and oil. Or so it is commonly claimed and generally believed.

Trouble with the biogenic theory of petroleum formation

There is, however, an active scientific minority that says the biogenic theory for the origin of petroleum is fundamentally incorrect and that the biomolecules found in

* Cyanobacteria, formerly called algae, are commonly referred to as pond scum.

petroleum are later additions, contaminants introduced by bacterial action, or biomaterials dissolved by nonbiogenic oil of deep origin during its migration toward the Earth's surface.

Among the several difficulties encountered by the conventional biogenic theory is the fact that nothing chemically resembling natural crude oil has ever been manufactured from plant material under conditions that might exist in nature.[3] Ordinary marine organic matter contains about 7 to 10 percent of the light element hydrogen, but the hydrogen content of natural petroleum is around 11 to 15 percent, and oil from some deep sources is even more rich in hydrogen. Where did the extra hydrogen come from? Deep oils may also lack all indications of biological activity.

Scientists who favor a nonbiological origin for petroleum hold that deep oils are rich in hydrogen when formed, but that bacteria strip them of some of their hydrogen, whether at depth or on the way toward the Earth's surface. Adherents of the conventional biogenic theory counter by emphasizing the time element, arguing that attempts to manufacture petroleum in laboratories are too rapid, producing "undercooked" mixtures poor in hydrogen.[4]

In 1944, Otto Schmidt (1891–1956), astronomer, mathematician, geophysicist, Arctic explorer, and statesman, proposed an alternative to the Fireball Earth theory.[5] Schmidt (**Fig. 2.1**) hypothesized that cosmic dust had clumped together in the turbulence of the nascent Solar System, eventually forming particles. The particles had clumped to become small balls, then larger balls, and so on eventually to form the planets. The process greatly accelerated once the bodies attained sizes of approximately one kilometer, at which point they attracted each other directly through their mutual gravity.

Figure 2.1 Soviet postage stamps honoring Otto Iulevich Schmidt.
http://www.stamprussia.com/487b.jpg (3 kopek)
http://www.matematycy.interklasa.pl/znaczki/schmidt/01.jpg (4 kopek)
http://www.shipstamps.co.uk/wp-content/uploads/10389/tmp15E.jpg (10 kopek)

Schmidt's views lead to a different sort of Earth, one that had never been entirely molten, because dust and small particles — whether cosmic or not — cannot retain heat. They are like the breakfast crumbs that rapidly cool to room temperature while the toast stays warm, a familiar phenomenon that is due to the scale of things, for small particles have large surface areas in relation to their tiny volumes, and this causes them to lose their heat more rapidly.

According to Schmidt's scenario, the Earth had necessarily been formed cold and despite subsequent melting here and there at different depths throughout geological time, it had never been a molten fireball.

In Schmidt's time, news traveled slowly between Moscow and Massachusetts where I lived and in any case, Schmidt was perhaps best known in the West for his position as a member of the Central Executive Committee of the USSR. Thus, it was not until the detailed follow-up work by the astronomer Viktor Sergeyevich Safronov (1917–1999), also of the USSR, that such ideas entered Western scientific thought, and it was not until the late 1980s that a coherent picture emerged in which "planetesimals," and then planets, accreted from a thin dusty disk of gas rotating about the Sun. In some accounts, the work of Schmidt is ignored and full credit is given to Safronov, whose *Evolution of the Protoplanetary Cloud* appeared in English in 1972. This is incongruous and a bit sad, given that in 1974 Viktor Safronov was awarded the Otto Schmidt USSR Academy of Sciences Prize.

From the 1950s or 1960s, the theorizing of scientists in Russia and the Ukraine was no longer constrained by the fireball hypothesis, and some Soviet scientists began to have a new look at the generally accepted theory for the origin of oil and gas, and to find numerous inconsistencies. In this they were supported by the known views of Dmitri Mendeleev (1834–1907), discoverer of the periodic nature of the elements, one of the great scientists of all times, and a Russian national hero. For as early as 1877, Mendeleev had set out reasons favoring a nonbiogenic origin, insisting that petroleum was "born in the depths of the Earth, and it is only there that we must seek its origin."[6] He further argued that there had never been much oxygen in the Earth's interior and in consequence, that the production of reduced forms of carbon such as petroleum should have been favored over the production of carbonates.

Although details have been much debated, the ideas of Schmidt and Safronov have now been broadly accepted for a couple of decades in both East and West. Yet reasoned arguments for a brief fireball period are again being heard. Most emphasize one or two items in the considerable list of possible sources of heat: heat given off by gravitational compression of dust particles as they accreted, heat produced by the "freezing" of the Earth's iron core, heat generated by impacts with meteorite-like bodies, and heat from the decay of now extinct radioactive isotopes, all factors that were consequential in the distant cosmological past. These are quantitative matters that involve numerical assumptions and calculations and for the time being, there is no consensus whether, taken together, they would have been sufficient to cause the entire Earth to become molten.

In the meantime, a Russian-Ukrainian "nonbiogenic," or "abiogenic," or "abiotic," or "abyssal" school of thought concerning the origin of petroleum had been born and acquired a life of its own. And in the meantime too, scientific understanding had advanced on numerous fronts. Most notable, perhaps, was the work in the mid-1960s of the Ukrainian scientist Emmanuil Bogdanovych Chekaliuk (1909–1990), who showed that under high temperatures and pressures, nonbiogenic petroleum could, in theory, at least, be generated from mantle rocks containing the iron-rich variety of the mineral olivine, and that once formed, petroleum in the Earth's mantle could remain chemically stable. (The process for the generation of petroleum at depth is termed "serpentinization" because one of the end products is a rock composed of the greenish minerals of the serpentine group.) Chekaliuk eventually became the head of an "Abyssal Hydrocarbons"* research unit in Lviv (Lvov), but his career was greatly hindered by the fact that his brother was a militant anti-Soviet Ukrainian nationalist. Yet it is apparently due to the contributions by Chekaliuk and his research unit that the abiotic theory became known as the "Russian-Ukrainian" theory rather than the "Russian" or "Soviet" theory.

Oil wells that refill

Scientists in the West who were aware of Viktor Safronov's writings and those of Emmanuil Chekaliuk included the astronomer Thomas Gold (1920–2004), long-term director of Cornell University's Center for Radiophysics and Space Research. In a number of general and scientific articles and the two books cited in the previous chapter,[7] Gold argued that the early Earth had not resembled a hot body that was cooling down. Instead it had had the characteristics of an initially cold body that was heating up, with volatile materials driven upward by their buoyancy wherever and whenever pathways were available.

Light materials rise and heavy materials sink and there is ample evidence that our planet is still "outgassing." The great quantities of CO_2 spewed out by volcanoes can be interpreted as indicating that the carbon outgassing from deep in the Earth is mostly oxidized. But magmas and lavas contain oxygen. So once entrained in a volcanic system, any carbon-containing gas, reduced or oxidized, would be in contact with oxygen "all the way up." In such settings, the production of CO_2 is favored over CH_4, whatever the starting material.

* Hydrocarbons, which constitute the essence of petroleum, are compounds composed of hydrogen and carbon only.

The depths involved may be substantial, perhaps 250 kilometers, for as Gold emphasized, magma-filled channels may be maintained open to great depths because magmas have almost the same density as the rocks that encase them.[8]

Very different upward routes exist through cracks in solid rock, via what P.N. Kropotkin and B.M. Valiaev in 1976 called "cold pathways."[9] In such situations, initially rich methane-containing fluids (by which is meant methane-rich mixtures of CH_4 and CO_2 in chemical equilibrium with one another) rise and may escape to the Earth's surface regions. They rise through perennially or sporadically open faults and may survive unoxidized because they rapidly consume all the oxygen present on the surfaces of the faults. When such fluids reach the Earth's surface, the methane is oxidized while heavier hydrocarbons, if any, form oil seeps.

In places the rising fluids encounter an impervious cap rock. They are then blocked and forced to spread out until they spill upward over the highest edge of the cap rock. (When discovered, many or all gas fields are in the process of spilling upward.) As additional methane percolates upward, polymerization occurs beneath the cap rock and produces complex hydrocarbon molecules, oil. In such settings, microbes metabolize whatever they can, and leave their biological overprint.

At the super-giant Romashkino oil field in Tatarstan in the Volga-Ural region of Russia, oil wells extend as far as they eye can see on a grid with 400-meter spacings. The wells are gently pumped, day and night, summer and winter, mud and dust, but a dozen or more wells, termed "anomalous," are quite different from others. For while most of the wells are being gradually depleted, the anomalous wells refill as they are pumped. When I asked about the reserves of such wells, the operating company's chief geologist hesitated and then said, "until the year 2100 plus."[10]

Vladimir A. Trofimov, a geologist who has extensively studied the Romashkino oil field, has obtained seismic profiles showing very deep, thin, vertical structures that he convincingly interprets as fracture channels.[11] According to Trofimov, oil fields are composed of three parts:

- An oil-filled trap,
- A deep [mantle] reservoir as a source of hydrocarbon fluids, and
- An oil-refilling [fracture] channel connecting the deep reservoir to the trap.[12]

The channels described by Trofimov are thin, linear, kilometers-long "tectonic faults and narrow zones of fractured rocks" that are nearly vertical in the local sediments and uppermost crystalline basement, but become inclined at deeper levels in the Earth's crust. Overall, Trofimov views oil fields as "complex and constantly active hydrodynamic systems," affected by many factors, even perhaps the tides.[13]

A 17-year monitoring project carried out at half-year intervals showed that the density and viscosity of the oil at Romashkino varied over the long term, suggesting periodic inflow of light materials.[14] On the whole, anomalous wells produced lower density petroleum than those that did not refill.[15] Trofimov saw channels as sporadically or constantly active, but flow was not simply on or off, and the isotopic composition of carbon from gas samples changed after small earthquakes.[16]

Ideas such as these are not entirely new. In 1976, a *review* article by Peter N. Kropotkin and B.M. Valiaev[17] concluded that "the scheme outlined here [is] for the formation of oil as a residual deposit of the degassing of the Earth's crust and mantle and condensation of heavier hydrocarbons from the primordial gas flow.... We cannot exclude still other models of inorganic synthesis of oil." The article commenced with a long list of references, mostly to Russian publications, beginning with the writings of Mendeleev, "on the foundations of the theory of the deep origin of oil."[18] Most of the references remain generally unfamiliar in the West.

Western oil companies are still very much tied to the notion that oil has an ultimately biological origin. Their best argument is that the biotic theory has enabled them to find oil and gas decade after decade. But those who favor an abiogenic origin point to the formerly unexpected places where methane plus a "surprisingly large number of molecules ... used in contemporary biochemistry on Earth" have now been reported.[19] Carbon-containing molecules have been detected in recent years in the interstellar medium, the atmospheres of all the outer planets, planetary surfaces, the satellites of the giant planets (in both the atmosphere and the surface of Saturn's moon Titan), comets, asteroids, meteorites, Kuiper Belt objects, and interplanetary dust particles.[20] This compilation indicates that complex carbon-containing molecules are easy to form and are most probably produced by numerous different chemical and cosmochemical pathways.

The debate has not been polite. Personal reputations, the price of oil, and the wealth of nations are all at stake. On first entering the field in 1976,[21] Tom Gold believed that many of the key ideas concerning the true origin of petroleum had originated with him. Yet, four years on, while working out the details (wherein resides the devil), he discovered the Russian-Ukrainian school in which top-notch scientists had been struggling with the same problems for a decade or two. Gold thereafter worked assiduously to give priority where it was due, paying to have some of the Russian-language publications translated into English and then distributing the translations without cost. But it was too late. One particularly hostile adversary accused Gold of knowing Russian (which was not so) and having plagiarized information published in Russian (also not so). At the same time, some American detractors discounted the Russian-Ukrainian work as pseudoscience (which was certainly not the case).[22] Similar disagreeableness is known in Russia too, where the Russian-Ukrainian abiogenic theory remains a minority position and where professors have been known to be harsh with doctoral students whose theses support the abiogenic theory.

My own view is that methane and more complex hydrocarbons are readily formed substances, that much methane and oil have been formed by nonbiological processes, and that this spectre must surely haunt certain individuals, faculties, and boardrooms.

Following further background material, I will give additional reasons for my position. In general, I am simply recognizing that the abiogenic theory provides a point of view from which much of geology (and much else too) appears "in its greatest simplicity." This was the goal famously set out by Willard Gibbs (1839–1903), perhaps the greatest American scientist of his day, "as one of the principal objects of theoretical research in any department of knowledge."

Theia and the giant impact hypothesis

The plot thickened considerably around 1984. Schmit, Safronov, and Gold were not shown to have been wrong: despite a certain amount of heating, as noted above, the Earth had indeed most probably been formed cold, just as they claimed. But it also turned out that the early Earth had become mostly or entirely molten shortly thereafter. "Theia" is the name of the responsible party.

Commencing with Safronov's work on planetary formation, William K. Hartmann and Donald R. Davis devised what is known as the giant impact hypothesis for the origin of the Moon.[23] According to Hartmann and Davis, the Moon came into being when our planet received a glancing blow or near-direct encounter with a Mars-sized object. This planetlike object was subsequently named after Theia, who in Greek mythology was the mother of the moon goddess Selene. The theory of Hartmann and Davis, formulated in the mid-1970s, "languished until 1984 when an international meeting was organized in Kona, Hawaii, about the origin of the moon. At that meeting, the giant impact hypothesis emerged as the leading hypothesis and has remained in that role ever since. Dr. Michael Drake, director of the University of Arizona's Planetary Science Department, recently described that meeting as perhaps the most successful in the history of planetary science."[24] The encounter with Theia would have melted the outer thousand kilometers of the Earth but it may not have entirely melted the entire Earth.[25]

According to the giant impact hypothesis, the Moon has always been close to the Earth. And since the Moon has been severely bombarded, Earth too must have been severely bombarded, for Moon and Earth are sufficiently close to one another to constitute a single target in space. (Before the giant impact hypothesis, "Moon capture" theories had been popular, along with various notions concerning the whereabouts of the Moon before meeting up with the Earth. In those times, a "co-bombardment" of Earth and Moon, if envisioned, would have made little sense.)

The Late Heavy Bombardment

When severely shocked, as by a meteorite impact, rocks give up certain trapped products of radioactivity accumulated over time. This "resets their clocks" at zero, making it possible to determine how long ago the shock-event occurred.

Laboratory determinations on impact-shocked Moon rocks recovered by the Apollo and Luna teams produced a pair of unanticipated results, a double surprise for the community of planetary scientists and interested geologists. The first surprise was that there was little spread among dates: the clocks of most of the lunar rocks had been reset during a single bombardment episode, rather than at various times over the course of the history of the Earth–Moon system. The second surprise was that the bombardment occurred a *very* long time ago, some 4100 to 3800 million years ago. By comparison, the oldest rock formations known on Earth are "only" a bit older than 3600 million years. See **Table 2.1**.

Table 2.1: Earth's early chronology.

Chronological landmarks	Comments	Dates (millions of years ago)
Age of the Earth	Formed cold, but followed by some heating	4540
Formation of the Moon	Some or total melting of the Earth and loss of all or some of its volatiles	4527
The Late Heavy Bombardment (LHB), a unique episode that scarred the inner planets	Delivery to Earth of volatiles from the outer Solar System; heating but no melting of the Earth's mantle	approximately 4100–3800
Radiometric ages obtained from parts of the possibly unique Acasta rock formation of northern Canada	Was the Acasta Formation a "survivor" of the LHB?	4031±3, 3960, 3580
Oldest of several rock formations with similar dates	Cosmology gives way to geology	3600+

With these various considerations in mind, calculations concerning the fate of the Earth during the bombardment episode were then made, extrapolations based on the numbers and sizes of the lunar impact craters, and it turned out that the Earth must have undergone an enormous pelting, the largest impacts of which would have produced craters up to perhaps 5000 kilometers across. The bombardment, a unique never-to-be-repeated episode, and subject of the next chapter, is known as the Late Heavy Bombardment, abbreviated LHB, where "Late" refers to the timing of the event within the formation of the Solar System.

Planets nearer the Sun were born enriched in elements such as iron, silicon, and aluminum, whose atoms condensed at high temperatures from the primordial dust cloud. In consequence, these planets, including Earth, are "rocky." The outer planets were formed in far colder circumstances and are for the most part composed of gases and ices. (Lakes and rivers on Saturn's moon Titan are composed of methane (CH_4) and ethane (C_2H_6), and Titan also has dunes of what may be methane sand.) Still farther from the Sun, the outer regions of the disk were primarily composed of light elements, hydrogen, carbon, and oxygen. These reacted with one another within the disk to form water and hydrocarbons. (Helium, also present in the outer reaches of the disk, is chemically inert.)

Carbon-rich meteorites called carbonaceous chondrites are perhaps the most informative type of meteorites, vestiges from the cold outer reaches of the Solar System of small bodies whose gravitational compression never produced significant melting.[26] Carbonaceous chondrites are said to be "primitive" because they have been relatively little altered since the birth of the Solar System, some 4567 million years ago.

The content of water in carbonaceous chondrites is high, up to 22 percent in one case,[27] and they also contain great varieties of hydrocarbons and of other carbon-containing molecules whose designation as "organic" does not imply or even hint at biological activity. Terms such as *organic molecule* and *organic chemistry* are historical vestiges, relics of a quasi-religious nature dating to times when it was believed that constituents of living matter could only come into being through the action of an imagined "life force." We are now stuck with the misleading word "organic," which survived Friedrich Wöhler's laboratory synthesis of urea in 1828 and the subsequent synthesis by others of fats and sugars and other molecules present in living creatures. These days the unfortunate term *organic molecule* simply denotes molecules that contain carbon* with no implication that animals, plants, or a "life force" might have been involved in their formation. Organic molecules that have been found in carbonaceous chondrites include materials akin to kerogen (seen by proponents the biotic theory as the precursor of petroleum), some amino acids (essential for the formation of proteins), polycyclic aromatic hydrocarbons, alcohols, and fullerenes.

Great numbers of impact craters are visible on all the inner planets, not just the Moon. Early attempts to explain this ubiquitous cratering included mathematical models designed to show how each surviving planet might have swept up the primordial debris orbiting the solar disk in its neighborhood. Yet no amount of mathematical juggling could bring these models to produce realistic results. Then, commencing around 2005, scientists initially working at the Nice Observatory in France developed a very different theory. The "Nice Model," discussed in the next chapter, showed that the objects composing the Late Heavy Bombardment were derived from the cold outer regions of the Solar System. This accounted for earlier isotope studies that had shown that terrestrial water must have been "imported from the coldest regions of the Solar System after the Earth had formed,"[28] a conclusion that is assuredly also valid for "organic" materials such as those in carbonaceous chondrites.

The origin of our planet's water is not the subject of this chapter. But for convincing arguments that Australia's Great Artesian Basin is recharged from below and not by surface rainfall, see the early chapters of *A Voyage of Discovery* by Lance Endersbee (1925–2009). Endersbee lamented that this recharge is grossly inadequate, and he went on to retell how when the artesian waters of the Great Artesian Basin were first discovered, "the gushing water brought with it great quantities of natural gas, which was wasted and regarded as a nuisance." At one stage about 1910, wrote Endersbee, "the Chief Geologist in Queensland noted that there was no consideration of the possibility that the artesian water and the natural gas may have come from the same source, and noted that nobody was looking for a surface intake for natural gas. He was disregarded as a cynic. The belief in rainfall recharge of the artesian waters held firm."[32] Endersbee's later chapters treat Dmitri Mendeleev, to whom *A Voyage of Discovery* is dedicated, mud volcanoes and their tendency to catch fire, and the work of Tom Gold.

To summarize, the Earth was partially, or perhaps entirely, melted in the cosmological past when part or all of its primordial stock of light materials was driven off. Our planet's crust and upper mantle subsequently acquired a new stock of light elements at the time of the Late Heavy Bombardment. Light materials rise and the outgassing of our planet continues. Fluids rise from the crust, from the upper mantle, and from water-rich, carbon-rich slabs of rock that were formerly located on the Earth's surface but were subsequently subducted (Chapter 1). Some of the outgassing materials may derive from still deeper sources. The contribution of biology is uncertain.

Ice meteorites?

The Condon Report on the *Scientific Study of Unidentified Flying Objects* (1968) found that "ice fell out of cloudless skies" and suggested that such "ice chunks may arise from electrical effects of bolides or (more probably) may be the meteorites themselves."[29] Water ice was subsequently detected on Jupiter's moon Europa and on the polar caps of Mars. In addition, a report on an object circling the Sun beyond the orbit of Neptune, in the far reaches of the Solar System, noted that "Like the surface of many outer Solar System bodies," the surface of Kuiper Belt Object 2003EL61 "is rich in crystalline water ice."[30]

Scientists prefer hard evidence and have been historically reluctant to accept eyewitness or "anecdotal" reports of "ice meteorites." Blocks of ice observed or found in unusual circumstances tend to melt away without being investigated, especially, I imagine, when they emit an organic odor, as has been reported now and then. So when erroneous observations or unusual hailstones cannot be invoked, it is commonly claimed that such blocks originated as the frozen waste from the toilets of airliners. Yet anecdotes from times before commercial aviation, and the detection of ices and organic molecules in various parts of the Solar System in the decades since the Condon Report, suggest that it is more prudent to assume that ice meteorites do exist than that they do not.[31]

---

* Elemental forms of carbon, including graphite, diamond, carbonado, lonsdaleite, and chaoite, are not classed as hydrocarbons.

Origins

By far the most common mineral in the Earth's upper mantle is olivine, a silicate of iron and magnesium. In a series of complex chemical reactions that can take place at relatively low temperatures,[33] varieties of olivine may produce minerals of the serpentine group. This multistep process is sufficiently common to merit a name, "serpentinization," mentioned earlier. During serpentinization, olivine is transformed to give

serpentine + hydrogen *with or without hydrocarbons.*

Intermediate reactions on the way to serpentinization involve water that infiltrates olivine along fractures and grain boundaries, producing magnetite, silica, and hydrogen:

iron-rich olivine + water → magnetite + aqueous silica + hydrogen

$$(3Fe_2SiO_4 + 2H_2O \rightarrow 2Fe_3O_4 + 3SiO_2 + 2H_2)$$

Hydrogen, a highly reactive gas produced in the course of serpentinization, may in turn combine with any available carbon to form methane and heavier hydrocarbons. This occurs under conditions in which amino acids and other the building blocks of life would be stable.[34] Hydrogen is also a suitable source of energy for microbes, deep energy for deep microbes.

Until recently, the importance of serpentinization had been little examined and the work of Chekaliuk greatly underrated. It was understood that such reactions made sense thermodynamically and that they *might* take place in the Earth's upper mantle, but few scientists were convinced that they actually did occur to a significant extent. Yet some may have had their suspicions. Writing in 1932, and clearly constrained by the belief that oil and gas were derived "from organic matter, largely of vegetable origin," Sidney Powers and Frederick G. Clapp surveyed the "Nature and origin of occurrences of oil, gas, and bitumen in igneous and metamorphic rocks," making references to serpentine in both their abstract and their first paragraph.[35]

In 1957, Petr N. Kropotkin (1910–1996) asked if oil originated from life or if life originated from oil,[38] and Tom Gold nicely rephrased the question, asking whether gas and oil represented "biology reworked by geology" (the conventional view) or whether they were primordial materials that had acquired a biological overprint ("geology reworked by biology"). The question was not new and does not even date to the 20th century: less than a decade after the drilling of the famous Drake Well (**Fig. 2.2**) in western Pennsylvania in 1859, debate already raged whether oil and gas were of organic or inorganic origin, primordial or of subsequent manufacture.

Avenues and streets all over France are named in honor of Marcellin Berthelot (1827–1907), one of the great chemists of the 19th century, in whose time many still believed it inherently impossible to produce fats and sugars and other complex carbon-containing molecules from inorganic substances. (Wöhler's production of urea in 1828 had been accidental and incidental, and with little immediate impact.[39]) Berthelot demonstrated once and for all that the processes of life can be explained by the laws of physics and chemistry and that they did not require the life-spark required by centuries-old notions of "vitalism." By 1866, less than a decade after the drilling of the Drake Well, Berthelot applied his thinking to natural petroleum, concluding that it was possible to "conceive the production by purely mineral means of all natural hydrocarbons. The intervention of heat, water and alkaline metals, [and] the tendency

Where did life originate?

One of the geological settings in which serpentinization occurs is at hot mid-oceanic vents. Sunlight does not reach such locales and hydrogen-based biological activity exists with extremely little[36] energy input from the Sun, whether through photosynthesis or any other means. It might be claimed that life began in such settings, a contention backed by the fact that serpentinization should have occurred at much shallower depths in the hotter Earth of 3 billion years ago.

Most or all of the multicelled creatures (metazoans) that live at such vents these days — specially adapted clams and crabs, for instance — have evolved from familiar deep- or even shallow-water creatures, and not locally at the vents themselves. Yet single-celled creatures are present in great abundance at the vents and in intergranular spaces and cracks down to a depth of several kilometers. It is perhaps at depth that life first came into existence, enabled by locally generated H_2 and CH_4. In that case, according to Gold, vents and similar locales would be halfway houses.

In his idiosyncratic *Cancer Selection* (1992), James Graham provided arguments that intriguingly embellish Gold's view, claiming that the earliest multicelled animals had been burrowing creatures, highly susceptible to cancers caused by ultraviolet radiation. According to Graham, eyes did not develop in order to see, but in order to avoid UV.[37] Single-celled creatures are not subject to cancer, but they too may be damaged by UV and cosmic radiation. This suggests that life may have originated in cracks and intergranular spaces, harnessing sources of energy available at depth, and using the upper several meters of the sea as a radiation shield.

of hydrocarbons to unite together to form the more complex materials, suffice to account for the formation of these curious compounds. Moreover, formation will be continuous because the reactions which started it are renewed incessantly."[40]

Figure 2.2 Woodford and Phillips wells, Tarr Farm, 1861. Courtesy of the Drake Well Museum, Pennsylvania Historical and Museum Commission

Eugene Coste (1859–1940), one of the fathers of the Canadian natural gas industry and an accomplished scientist,[41] also preferred the nonbiogenic origin for natural gas. According to Coste, "Oil, gas and bitumens are never indigenous to the strata in which they are found — they are secondary products impregnating and cutting porous rocks of all ages, exactly as volcanic products alone can do" ... "volcanic emanations condensed and held in their passage upward in the porous tanks of all ages of the crust of the earth from the Archaean rocks to the Quaternary."[42]

And Hollis Hedberg (1903–1988), seen by many as the greatest petroleum geologist of his times, was clearly troubled by the suspicion that "something was wrong." Hedberg was also much intrigued by the origin of coal. Some black coals* are so extremely *chemically* pure that it seems they could have only been formed by deposition from a fluid. And many coal seams are also *physically* pure, lacking the soil and debris that one would expect. Such facts require explanations.

Sir Robert Robinson (1886–1975), Nobel Laureate in chemistry in 1947, was perhaps the first to argue in a scientific journal that the origin of petroleum did not have a singlet explanation. Petroleum, he argued, was a material of inorganic origin to which molecules of biological origin had been added.[43] Robinson wrote of the "duplex origin of petroleum" and he was surely correct, unless the origin was actually "triplex" or "multiplex." For, as suggested by the widespread occurrences throughout the Solar System and beyond, hydrocarbons are easy to make.

In the mid-20th century, Mendeleev's ideas were "revived and further developed by N.A. Kudryavtsev, P.N. Kropotkin, V.B. Porfiriev and other scientists of the former USSR."[44] Kropotkin placed the matter in the context of "Degassing of the Earth,"[45] of heavy materials sinking and light materials rising. Eight international and Russian conferences (1976, 1985, 1991, 2002, 2006, 2008, 2010, 2013) were held on these problems and five collections of scientific articles were published (1980, 2002, 2006, 2011, 2013).[46]

Nikolai Alexandrovich Kudryavtsev (1893–1971), who was cited just above, was an extraordinarily successful oil explorationist, largely responsible for the discoveries of commercial oil fields in a half dozen regions of the Soviet Union. Kudryavtsev's scientific approach was exceptional for his times too. For whereas oilmen, East and West, had by then universally accepted the biogenic theory of petroleum formation and had successfully

* Peat and some brown coals are excluded here and throughout my discussion.

discovered sufficient reserves to supply the world's rapidly growing energy requirements, Kudryavtsev ardently advocated an abiogenic origin for petroleum.

Despite his great success as an explorationist, Kudryavtsev spent time in the Gulag as a consequence of his "deviant ideas." Professional jealousy may have been involved, but it appears that Stalin (d. 1953) had seen Kudryavtsev's successes as tarnishing the reputation of the greatly revered Russian polymath M.V. Lomonosov (1711–1761), credited in Russia with originating the idea that oil came from the "percolation" of biological material. (The full name of Moscow University is "Lomonosov Moscow State University.")

Modern abiogenic theory is born

With the formulation of what has become known as Kudryavtsev's Rule in the 1950s, the modern abiogenic theory was launched. This rule states that if hydrocarbons — gas, oil, or coal — are found at one level in any particular area, they will also be present in large or small quantities at all levels all the way down and into the local basement bedrock. The rule is valid, whatever the type of rock, but whether the occurrences at any given level are of commercial value depends on the porosity and permeability of the local rocks.

Kudryavtsev also gave a long list of oil fields under which, he claimed, vertical faults could be identified or surmised.

Kudryavtsev's presentations and examples were not limited to oil fields in the USSR. In one of many cases, he noted that oil had been obtained from the basement up through each unit of overlying sedimentary rocks at the Lost Soldier Field in Wyoming. Oilmen conventionally speak of "source rocks" and "reservoirs," a matter that commonly implicates horizontal flow. Kudryavtsev, however, argued that oil and gas are found wherever permeable rocks are overlain and capped by impermeable rocks, thereby evoking vertical flow. On the need for an impermeable caprock, both approaches, biological and abiological, are in agreement.

Another of Kudryavtsev's examples was the great 6000-kilometer arc of the Indonesian islands into Burma, with oil and gas production, volcanism, and earthquakes in many places. Adherents of the conventional biogenic theory for the origin of petroleum fluids claim that such arcs correspond to ancient and present-day collisions between tectonic plates and are places where oil, gas, and water have been squeezed out of descending slabs of sedimentary rocks. This is correct. But when one plate passes beneath another, it also rejuvenates ancient fractures and opens access to deeper domains.

In more recent years, Irina N. Plotnikova has reported commercial quantities of oil or gas from "crystalline basement rocks" from more than 300 deposits on different continents. ("Crystalline" in such cases refers to metamorphic and igneous rocks; unless they are fractured, crystalline rocks have far less porosity than sedimentary rocks such as sandstones.) Plotnikova, too, favors vertical outgassing, and has calculated that the sedimentary rocks claimed to have been the source-rocks of Tartarstan's Romashkino super-giant oil field, cannot possibly have held as much petroleum as has already been produced. The oil must come from elsewhere.[47]

In earlier times it would have seemed that a suitable petroleum source-rock was available for the great oil fields of Saudi Arabia. Carbon-containing sedimentary rocks were present in abundance and had been mapped in 3-D, and calculations indicated that a great deal of petroleum production could be expected. That was decades ago. In the meantime, production has continued and continued and has *very* greatly exceeded the earlier calculations. On rerunning the calculations, Kutcherov and Krayushkin (2010) wound up asking, "Where did 94% of Saudi Arabia's recoverable oil come from?" They then made similar calculations for the oil and bitumen belt in Alberta and Saskatchewan, and for Venezuela's Bolivar coastal oil field. The numbers do not work out. Something is wrong and the clear explanation is that these and other oil fields are being refilled while (and where) nobody is watching.[48] (Endersbee had reached a similar conclusion concerning the amount of water recovered from Australia's Great Artesian Basin.)

Similarly, mud volcanoes in China's Xinjiang Province (**Fig. 2.3**), in the Baku region of Azerbaijan, in Romania, in Indonesia, and elsewhere have liberated quantities of methane greater than the calculated reserves of any producing gas field. Decades ago Kudryavtsev asked where all this methane came from.[49]

The calculations for the petroleum reserves of Saudi Arabia are actually still more puzzling for those who favor a biotic origin. For oil and gas are also found in Bahrain, Iran, Iraq, Kuwait, Oman, Qatar, Syria, the United Arab Emirates, and Yemen, and the detailed chemical and trace-element signatures of these oils indicate that they all had the same source.[50] So the 94 percent problem is probably more like a 99 percent problem.

Figure 2.3 Summit of a mud volcano just outside of Kuitun, Xinjiang, western China, May 2010. A few years earlier, an explosion had awakened the residents of Kuitun, who saw a large vertical flame issuing from the area in this photo. Photo: J. Saul

Another sort of support for the abiogenic theory of petroleum formation, and against the horizontal migration of petroleum, comes from observations by a geologist whose job "has to do with cleaning up hydrocarbons spilled in the subsurface," who writes, "I know from my work that most compounds won't be transported more than a thousand feet because bacteria will eat them and they will disperse. Since I have learned how petroleum moves in the subsurface, transportation from source material to traps seems pretty far-fetched to me. I tend to now favor the abiotic theory."[51]

Oil and gas from particular regions — the Middle East, Panhandle, North Sea —have distinctive chemical signatures. There is nothing subtle about this: petroleum from diverse regions is sufficiently distinctive to require separate pricing. To my mind, this suggests great reserves at great depths and evokes a "plums-in-a-pudding" structure for the upper Earth. Each plum might correspond to one of Safronov's planet-forming "planetesimals," each with its distinctive chemistry. Or, more likely, each plum might be due to an impacting body that had arrived from the outer reaches of the Solar System during the Late Heavy Bombardment, each with its distinctive chemistry.

Methane (which has one atom of carbon), ethane (two carbon atoms), and propane (three) are the first members of the "alkane" series of chainlike molecules, the lighter members of which can be produced by microbes.[52] At the Rainbow site (36°15′N) on the mid-Atlantic Ridge, alkanes with 11 to 22 carbon atoms have been found but, according to Kutcherov and Krayushkin, "Contemporary science does not yet know any microbe which really generates" such molecules (and similarly for other complex and branched hydrocarbons).[53] To this can be added recent analyses of gases emitted from La Fossa Crater on the island of Vulcano (Aeolian Islands, Italy) from which traces of more than 200 different carbon-containing compounds have been identified.[54]

Kutcherov and Krayushkin also cite six particular occurrences of oil and gas in crystalline rock, presumably also included among the 300 repertoried by Plotnikova, claiming that no nearby, or even distant, sedimentary rocks can have been their sources. These occurrences "cannot be explained from the traditional biotic petroleum origin point of view," write Kutcherov and Krayushkin, who conclude that their source must be in the Earth's mantle.[55] One of their examples is the Volodarsk-Volynskii (Korosten) area in the Ukraine, famous for its large transparent gem beryl, topaz, and quartz crystals (Chapter 1). The area is also source of "kerite," an "organic material found during pegmatite mining ... as black fibrous aggregates of short, curving interlaced fibers of average diameter 14.7 microns" (**Fig. 2.4**). The few scientists who have studied this obscure material have failed to agree whether it is a primary mineral-like substance or the fossilized remains of cyanobacterial

Figure 2.4 Kerite fibers, Volodarsk-Volynskii (Korosten), Zhytomyr Oblast, Ukraine. Field of view approximately 4 mm. (The kerite is found in a pegmatite within a rapakivi granite, for which see Chapter 1.) John A. Jaszczak specimen and photograph.

mats. The kerite debate is a microcosmic version the biotic vs. abiotic argument followed throughout this chapter.[56]

In a section of their text subtitled "Petroleum fluid inclusions in minerals of igneous and other crystalline rocks," Kutcherov and Krayushkin record the presence of the light alkanes methane, ethane, propane, and butane within inclusions in quartz from the same mineral locality, Volodarsk (Korosten), and take this as evidence in favor of the abiogenic theory.[57] They give similar examples from all continents, including Antarctica, and end by concluding that the primary inclusions of petroleum fluids had formed in conditions present in the Earth's mantle, not its crust. ("Primary" in such cases refers to inclusions formed at the same time as the enclosing mineral itself, not formed later or introduced secondarily through cracks or fissures.)

Diamonds

Elsewhere Kutcherov and Krayushkin treat hydrocarbons and bituminous inclusions in diamonds. Diamonds, a type of pure carbon, can apparently crystallize quite readily by any of several much-debated mechanisms at depths of 150 to 250 kilometers or more below the Earth's surface.[58] Formed at these depths, diamonds cannot have reached the Earth's surface by any familiar means, by some sort of witnessed volcanic eruption, for example, or by the slow erosional removal of the many tens of kilometers of overlying rocks. In such cases, the diamonds would have been exposed to conditions of pressure and temperature in which they were unstable, causing them to degrade to graphite, perhaps even to volatilize.

Most diamonds were brought to the Earth's surface during eruptions of an unusual type of rock called "kimberlite," whose eruptions have never been directly witnessed. Kimberlite eruptions are violent gas-driven affairs with rocks and debris moving at supersonic speed by the time they reach the Earth's surface. Kimberlites may rise in the atmosphere (like an oil gusher, but far higher) and then fall back, forming a pile of debris upon a carrot-shaped pipe of kimberlitic rock that pinches out and disappears at depth. In addition to the minerals of deep origin used to define kimberlites, the pipes and the fall-back debris also contain fragments ripped off the walls of the surrounding rock and, in their uppermost sections, kimberlite pipes may even contain fragments of unburnt wood.

The rapidity of the ascent enables entrained diamonds to survive passage through the great vertical zone throughout which they are unstable and would rapidly degrade. Arriving near the Earth's surface, the gases that drive the eruptions expand many fold, and gas expansion is necessarily accompanied by cooling, as when air escapes from the valve of a tire. Cooling might amount to as much as 90° to 140°C (160° to 290°F), bringing the temperature of the kimberlite down below the kindling point of wood.[59] This too enhances the chances of survival for the diamonds. Even so, many arrive at the surface with resorbed or corroded surfaces (**Fig. 2.5**).

But what is the gas that drives the kimberlites? Carbon dioxide, which is odorless, is surely present, but in the diamondiferous provinces near Arkhangelsk, and in Yakutia, "diamonds are sought by smell like truffles," according to a Russian report of kimberlites containing certain diagnostic hydrocarbons in concentrations rarely encountered even in oil and gas deposits.[60]

Clues to the formation of deep abiogenic hydrocarbons

Evidence and arguments concerning deep abiogenic hydrocarbons come in many forms and, as others have discovered before me, the subject is most easily surveyed as a series of short listlike paragraphs. Some involve the origin of coal.

Figure 2.5 Resorbed diamonds, Crater of Diamonds State Park, Pike County (near Murfreesboro), Arkansas, USA. Sizes range from 7 mm (2.01 carats) to 13 mm (8.81 carats). These specimens seem to have been selected to illustrate the range of colors encountered in the Arkansas occurrence. Houran Collection, Joe Budd photo.

"It is interesting to note that diamonds from the Crater are all well rounded, a testament to the chemically corrosive conditions they encountered during their trip from the upper mantle to the surface. What typically are called crystal faces are actually absorption features, readily recognized by their curved faces and curved lines at face joins. The diamonds have lost approximately 60 to 80 percent of their weight during transport, due to this process" (Howard and Hanson, 2008, with thanks to Glenn Worthington).

Only about one kimberlite pipe in ten contains diamonds. It is unclear whether the other pipes had always been barren, or whether their diamonds had been totally resorbed.

- Tom Gold thought that most or all black coals originated as a nonbiogenic fluid, the Albertite Coal of New Brunswick, Canada, for example, which appears to have been injected under pressure into the surrounding rocks.[61]
- The Chimaera gas-seep near Olympos in southwestern Turkey emerges from a fracture along which limestone and serpentine-rich rocks are in contact. The gas has been burning continuously since the times, some 2500 or more years ago, when Hephaestus, smith of the gods, presided over the site. Detailed work at the site suggests that the gas, which includes methane, ethane, hydrogen, carbon dioxide, helium, and nitrogen, is a mix of both biogenic and serpentine-derived gases, and that the methane fraction, at least, is not of mantle origin.[62]
- A group of organic compounds called porphyrins includes heme, the iron-containing pigment in red blood cells. Iron and magnesium porphyrins are found in various plants or animals, but not in oils. By contrast, nickel and vanadium porphyrins, found in petroleum, are not present in plants.*
- Kutcherov and Krayushkin also surveyed the numerous meteorite impact-craters in which petroleum, whether oil or gas, commercial or noncommercial, has been found. They conclude that oil and gas in many such sites cannot be biogenic because inter-crater source rocks would have been destroyed or melted by the impact, hence, that the oil and gas must have come up from depths. Their logic is particularly convincing for sites in which the impact occurred in nonsedimentary, crystalline, rocks. In those cases, as in others, porosity and permeability are provided by fracturing produced by the impact.[63] See **Figure 2.6**.

* In East Africa, several varieties of transparent gemstones owe their attractive colors to minor or trace amounts of vanadium, and are found in host rocks that are rich in graphite and have an "organic" smell; see Chapter 1.

Figure 2.6 Flame from burning gas from a farmer's water well, Siljan meteorite crater, Siljan Ring, Dalarna, Sweden. There are no sedimentary rocks on or adjacent to the farm, only fractured granite. Photo: J. Saul.

- Many "organic" molecules — with the two sugars dextrose and levulose as familiar examples — have mirror-image forms, much like a pair of gloves, a phenomenon first described by Louis Pasteur (1822–1895). Biological processes systematically favor one of these forms over the other. Russian researchers have shown that petroleum of deep origin and synthetic (laboratory-made) petroleum never exposed to microbial action have the same mix of right- and left-handed molecules. But in petroleum of less deep origin, one form dominates over the other, a fact readily interpreted as due to microbial activity or contamination. Advocates of the conventional biogenic theory counter that the balance between right- and left-handed molecules in deep petroleum is a secondary feature, a resetting caused by "cooking" over long periods of time.

"Cooking" arguments based on the time required to produce petroleum are difficult to rebut because they invoke a time factor that is impossible to observe or to duplicate in the laboratory.

- Carbon exists in two stable forms, carbon-12 and carbon-13.* Photosynthesis favors the lighter isotope carbon-12 over carbon-13. Yet the ratio of carbon-12 to carbon-13 in oils of various ages is more or less the same in all samples, including those from periods when there was a great proliferation of vegetation. This suggests that oil is not derived from vegetation.
- Methane from the deepest sources, where microbes are few or absent, is depleted in carbon-13 which, being heavier than carbon-12, had apparently been slowed on its way up.
- The light, hydrogen-rich, components of petroleum are preferred by bacteria, but petroleum of deep origin is systematically rich in hydrogen. The apparent excess of hydrogen in deep petroleum may indicate that microbes had not yet had time to peel off (metabolize, eat) much of the original hydrogen.
- "Surface soils above gas fields have very high methane content," suggesting flow rates sufficient to exhaust the known reserves of any any gas field within a few thousand years. This is easy to understand if the fields are replenished from much larger reservoirs.[64]
- A similar situation exists as concerns the great quantities of methane continually released from coal beds.[65]
- Petroleum reservoirs refill spontaneously. An item by William R. Corliss (1926–2011), who compiled scientific anomalies of various sorts, reads: "Gulf of Mexico, Eugene Island 330 ... produced 15,000 barrels per day in the early 1970s. By 1989, the flow had dropped to 4,000 barrels per day. Suddenly, oil production rose to 13,000 barrels per day.... Surprisingly, the new oil was of different age and characteristics. This phenomenon is not uncommon worldwide."[66]
- Refilling, a well-known phenomenon, is due to fluid migration. Although horizontal migration is favored by most oilmen, Trofimov, who was cited earlier, argues convincingly for the vertical refill of the anomalous wells in Tartarstan's Romashkino oil field.[67]
- Helium is recovered from some gas wells, but always in association with methane, or with nitrogen with which high concentrations of helium have been found in a few gas fields. Helium is a product of the decay of uranium and thorium, a slow process that causes it to gradually accumulate in rocks. Yet high helium concentrations correlate poorly with local occurrences of uranium. The correlation is much better with the

* Carbon-14, which is not stable, is formed in the atmosphere by the action of cosmic rays on nitrogen.

depth of origin of local oil and gas. Gold, as summarized by Robert Ehrlich in his highly readable *Nine Crazy Ideas in Science*, explains this in the following manner: "The quantities of helium that build up in rock fissures are not great enough for the gas to attain enough pressure to reach the surface on its own — hence no pure helium gas is released. However, when methane flows through porous [or fractured] rocks, it can sweep helium along with it. Because helium is being created continuously from radioactive decay, more will be swept up by methane, the greater the depth from which methane originates."[68] The association with nitrogen is puzzling: it might indicate that nitrogen of particularly deep origin has swept up helium over longer vertical pathways than has the methane. The presence of helium deep in the Earth is a subject of much ongoing research.

- Volcanoes release great quantities of oxidized carbon, CO_2. But sulfur crystals from Mt. Etna and elsewhere in Sicily (**Fig. 2.7**) contain inclusions of petroleum and bitumen, that is, of unoxidized carbon.[69]
- In Kamchatka, in the Russian Far East, hot springs and volcanoes spew out small quantities of astonishingly unexpected "organic" materials: chloroform, carbon tetrachloride, and even CFCs ("Freon"),[70] thought to be partly responsible for the hole in the ozone layer.
- Particular carbon-containing molecules found in oil are used as arguments in favor of the biogenic theory. These same molecules are also found in coal. But petroleum is supposed to come from the decay of microscopic marine organisms, while coal is supposed to derive from the breakdown of land vegetation, and the logic and line of reasoning seem odd.

Figure 2.7 Sulfur with inclusions of petroleum, Cozzo Disi Mine, Casteltermini, Agrigento Province, Sicily, Italy. The largest crystal is approximately 5 cm in greatest dimension. Arkenstone Specimen / Joe Budd Photography

- In some black coals, wood fossils are entirely filled with carbon without being deformed, even at the cellular level. Such fossils could have been formed if the coalified wood had been invaded by a carbon-rich fluid much as petrified wood is formed by the invasion of a silica-rich fluid.
- Many coal seams, some of them meters thick, are essentially free of soil and other debris. How can that be? An answer, when one is obtained, is that the seams were formed by the decomposition of a floating mass of vegetation.
- The element germanium is 10,000 times more concentrated in certain coals than in the surrounding sediments. It is difficult to understand how this could have come about unless the germanium had been in solution at some stage or tied up in an "organo-complex" molecule or ion that had broken down at the pressure and temperature conditions at which the coal was formed. The elements mercury, uranium, and gallium, also found in trace amounts in coals and soils, are similarly concentrated in the coal but to lesser degrees. Bill Corliss distilled these facts from Gold's *Power from the Earth*. Corliss went on to make a point of his own, noting that there is an overabundance of carbon in coal (as compared with carbon in vegetation and peat) and that an external source of carbon seems necessary, as well as an external source for the germanium, mercury, uranium, and gallium.[71]
- Fossil plants and leaves are occasionally found in coal. This has been taken as evidence for the biotic origin of coal. Gold disagreed, countering that "if coal really is formed from the compression of stuff like plants, why should whole leaves occasionally survive intact? Indeed, finding leaf fossils in coal deposits is like finding a whole tomato in a tin of puree and insisting that the puree was made in a liquidizer."[72]
- Ordinary coal shows few indications of fossils. Many of the splendid coalified specimens in museum collections "come from rocks that are as much as 100 ft above or below any workable coal seam."[73] Here too, a carbon-rich fluid would seem to be implicated.

- The presentation of Kutcherov and Krayushkin ended with a section on the world's great reserves of natural gas hydrates in ocean-bed sediments, in Lake Baikal, in the Caspian Sea, and in permafrost. Such hydrates, which form under particular combinations of pressure and cold temperature, are composed of ice structures that have physically trapped molecules of gas without forming chemical bonds. Different types of gases may be trapped, including helium, argon, carbon monoxide, carbon dioxide, nitrogen, or light alkanes, especially methane. Methane and ethane hydrates produce "ice that burns." Kutcherov and Krayushkin argue that the source material of the world's reserves of gas hydrates and the underlying free gas is not organic and indeed, various studies of microbes in hydrate deposits have found few methane-producing species.
- Earlier, I noted that when discovered, many or all gas fields are overfull, spilling upward, presumably because they are in the process of being refilled. The conventional theory for the origin of oil and gas would have the refilling occur more or less horizontally from a source rock. But recent observations indicate that "extensive methane venting" is taking place in the Blake Ridge gas hydrate province on the southeastern U.S. Atlantic margin.[74] This leads to the suspicion that gas hydrate provinces are also refilling. But in the case of gas hydrates, it is far more difficult to argue for refilling by horizontal flow.

Deep gas and bubbles

At the time Gold's version of the deep gas hypothesis was formulated, it rested on two assumptions: first, that hydrocarbon molecules can exist at great depths, that is, that they are thermodynamically stable under great pressure, and second, that porosity of some sort — familiar or not — can exist at depth.

The first assumption no longer represents a problem. The work of Chekaliuk and his colleagues commencing 1968, cited in Gold's *Deep Hot Biosphere* (1999, Ch. 3), showed that methane and heavier hydrocarbons, including those found in petroleum, can exist in nonvolcanic regions at depths as great as 300 to 600 kilometers, and that they are in fact generated under such conditions. And by the 1980s or earlier, methane had been identified as a common microscopic inclusion in diamonds and other minerals that crystallize at great depths. Diverse high-pressure experiments have also been carried out over the decades to show how hydrocarbons might be synthesized by reactions among inorganic mixtures at pressures and temperatures at various depths in the Earth. Results have been promising but not fully convincing because of the possibility that air may have entered the mix.[75]

According to Gold and Soter,[76] methane and other fluids of nonbiogenic origin rise toward the Earth's surface in such a way that one "bubble domain" rises below another, somewhat akin to the strings of bubbles that rise in a glass of champagne.[77] Several steps are involved in their argument. First there is a gradual molecule-by-molecule accumulation of fluids by diffusion within the Earth's mantle. Then, as the accumulation grows in size, a vertical zone of interconnected fluid-filled pores may be produced. Buoyancy eventually causes the interconnected zone to rise and as it does, it encounters ancient fractures that are reopened and held open by the pressure of the fluid itself. The rising fluid manufactures its own "crack-porosity."

The reasoning depends on the fact that low-density fluid from a deep source cannot rise up to the Earth's surface as a continuous stream. Although this is intuitively clear, it merits an explanation.

The encasing rock is far more dense than the fluid. Thus, on passing downward through any particular vertical distance, the difference in pressure between the rock and the fluid must increase until the pressure of the rock eventually squeezes the bubble closed at what Gold and Soter termed "the critical layer." (They estimated that for methane, a "bubble-domain height" might be about 3 to 10 kilometers.) As the bubble continued to grow, the ever-increasing fluid pressure at its top would force new space to open. Opening would take place at the top, and closing would occur at the bottom. The new space opened up at the top of the bubble domain would mostly or only occur where the rising fluid encountered preexisting fractures and faults.

When gas or oil was pumped from the uppermost bubble domain, or when it leaked into the atmosphere or seeped onto the surface, the pressure difference between the topmost domain and the next underlying domain would increase. With less pressure to retain it, gas and oil in the lower domain could then rise, refilling the reservoirs from below.[78] The main objections to Gold and Soter's gas-domain theory seem to be quantitative, calculations and claims that the buoyancy of gas domains would never become sufficiently strong to overcome various resisting forces. At least one set of such calculations, however, considered a typical strength for the encasing rock, and failed to take into account the existence of perennially weak deep fractures.

Earthquakes are systematically associated with faults. And the scenario set out by Gold and Soter suggests that the rise of deep gas along faults might be the cause of some earthquakes.

Gold and Soter relate eyewitness accounts of earthquakes in which the rise of gas certainly appears to have played a role. In Norcia and Aquila (Italy), on 14 January and 2 February 1703, for example, "…the earth was here and there observed to split in cracks, from which streamed the evil odors of sulfur and bitumen; and men in Aquila most worthy of trust write that in many places after the earthquake sulfur and fire issued from the open earth." And commenting on an earthquake he witnessed at Cumana, Venezuela, on 14 December 1797, Alexander von Humbolt referred without elaboration to "earthquake flames" on the banks of the river Manzanares, as though it were a well-known phenomenon. Similar flames charred branches that overlay fissures in the ground in Sonora, Mexico, during the earthquake of 3 May 1887. I have intentionally cited old earthquake records here rather than more detailed recent accounts to discourage the commonly heard explanations of the flames and odors as due to broken gas mains or short circuits.

In 1995 Soter was fortuitously able to interview local people in the days following the June 15 magnitude 6.2 Aigion earthquake in the western Gulf of Corinth, Greece. People with whom he spoke reported "bubbling of the sea, extraordinary behavior of animals, earthquake lights, ground deformation, and the sound of wind immediately before the shock," all of which "might be understood as symptoms of gas venting before an earthquake." Soter also reported that sonar observations "made in 1988 showed a chain of submarine pockmarks tracing the Aigion Fault and evidently produced by the expulsion of fluids under pressure."[79]

Rising gas, oil, or water might lubricate or inflate existing faults, reducing friction and facilitating small tremors that could set off large earthquakes. But another mechanism might be responsible for violent quakes that occur with little warning and no evident trigger: a rapidly rising, rapidly expanding, gas domain might prop apart the sides of vulnerable faults, instantaneously reducing friction to near zero.

Soter also points out that "gas itself cannot supply the energy of an earthquake, but merely acts to trigger it, releasing the tectonic stress that has independently accumulated across a fault. In this view, *it is not an increase in rock stress* [as is broadly believed among geologists and geophysicists involved in earthquake research], *but a decrease in fault strength, that triggers an earthquake*" [Soter's italics].[81]

Use of anecdotal reports

Although scientists have a commonly justified adversity to basing conclusions on nonscientific or anecdotal reports, Soter showed how "multiple independent reports, even of low intrinsic probability" may be combined to increase the overall probability that a bizarre-sounding report is of scientific value. His method involves assigning values to (1) the probability that a report is accurate, and (2) the probability that it is relevant. The method is of broad applicability — to reports of ice meteorites, for example — but as regards earthquakes in particular the "accuracy of the report depends on the observational acuity, memory, and objectivity of the informant. The relevance of the phenomenon depends on its rarity, its proximity in time to the earthquake, and its similarity to precursors reported for other earthquakes."[80]

Investigators working in eastern Taiwan have observed a correlation between typhoons and the occurrences of "slow earthquakes," which release their energy over a period of hours to months. Some mechanism causes the low barometric pressure to trigger seismic activity. (Rain was eliminated as the cause.) The effect is small, though greater than that caused by tides, and was interpreted as producing "a very small unclamping" of faults that "must be close to the failure condition."[82]

This correlation of seismic activity with barometric pressure may simply be a matter of differences in gas pressures. Bubbles do not rise in a corked bottle of champagne, and the

Earth's uppermost rocks plus the Earth's atmosphere may be compared to a two-level "terra cork," restraining the rise of gas from below. The fractured, uppermost rocks correspond to a leaky cork, while the corking effect of the atmosphere is inherently weak. Lowering the barometric pressure might allow gas to seep, flow, or pop upward. In the case of the slow earthquakes of eastern Taiwan, it appears to seep. Before more violent earthquakes elsewhere, a rapidly rising or leaking gas bubble might saturate the soil, generate changes in well waters, pump radon from the soil,[83] produce odors and odd animal behavior,[84] and in certain circumstances, cause humans to feel unwell.

Accounts of flames and smells accompanying earthquakes in Italy, Venezuela, and Mexico were cited above. Hundreds of additional such reports have been registered here and there through the centuries.[85] In addition, there are many reports of peculiar behavior on the part of animals before, during, and after earthquakes, many involving burrowing creatures or small animals with their noses close to the ground.[86] A half dozen instances in the course of a single earthquake involve the deaths or pathological reactions of canaries,[87] well known for their susceptibility to methane and carbon monoxide.

The bubble-domain hypothesis may also shed light on other matters:

- The "overpressured gas" of the Gulf of Mexico and elsewhere may be gas rising from the next bubble in the bubble string.
- A gas domain might be blocked in its rise below a continent. This might occur beneath particularly thick, little-fractured parts of the Earth's ancient continental cores. Diamond-bearing kimberlites are systematically located in precisely such geological settings,[88] but the circumstances that enabled kimberlite eruptions to pierce 150 to 250 kilometers of crust are not clear. Gold and Soter suggest that these conditions might have come about when a gas domain, blocked in its rise, grew to great size and was then abruptly joined by the arrival from below of a second domain, the two domains interconnecting and coalescing, generating enormous buoyancy and blasting upward.[89]

There are places where gas, whatever its origin, seeps gently upward and can be "caught in the act," for example, at Siljan in Sweden (**Fig. 2.6**). In other places gas, whatever its origin, bubbles energetically upward, and can be photographed doing so, as it does just offshore at Mocha Island (**Fig. 2.8**) on the coast of Chile, where the Nazca plate is subducted under the South American plate. But visual evidence for Gold and Soter's deep bubble domains is not available.

Figure 2.8 Bubbles rising from the sea near Mocha Island, Chile, photographed from the shore by Jesse H. Ausubel, 12 January 2009.

A theoretical sketch by Gold and Soter of the two uppermost gas domains shows a highly irregular hypothetical shape (**Fig. 2.9a**). This irregularity probably expresses their intention to show that gas would not be able to impose a smooth bubble-form on its encasing rocks, which would necessarily have slight variations in permeability and strength here and there. (A champagne bottle made of weak, irregular plastic would similarly bulge here and there.) Such irregularities would be far less marked, however, if the bubble were not of gas, but of a magma whose density was much closer to that of the encasing rock. High-tech imagery of two magma domains beneath volcanic Mount Changbai/Paektu on the border of China and North Korea has recently been obtained (**Fig. 2.9b**), the lower of the two domains showing a regular bubble-form.[90] Mount Changbai/Paektu is located in an area where the Earth's crust may be particularly weak, a subject treated in the next chapter.

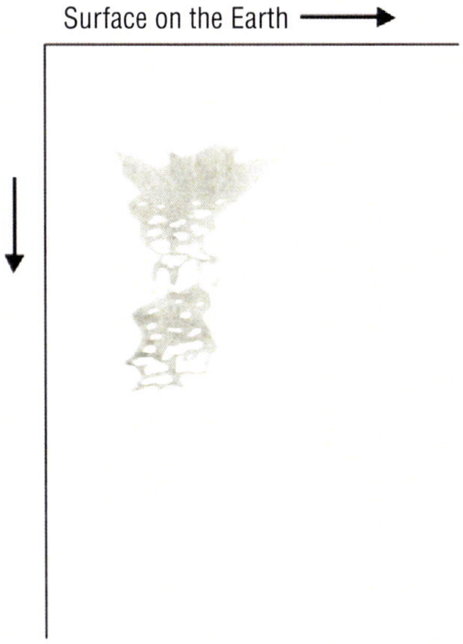

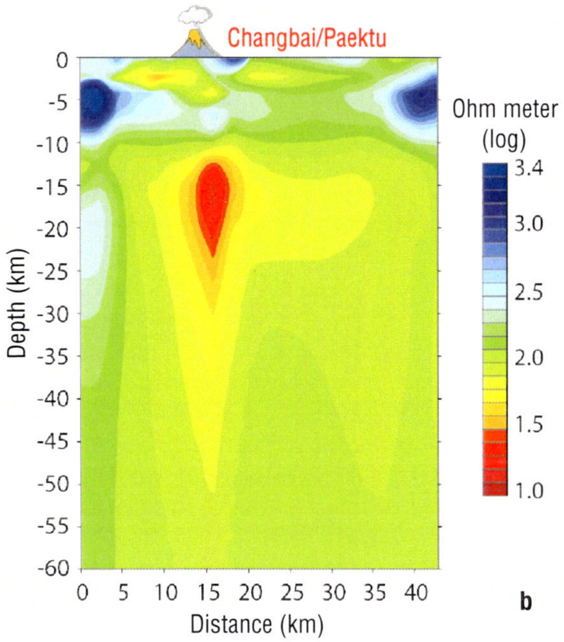

Figure 2.9a Theoretical cross section of two vertically adjacent fluid-filled fracture domains. The shaded areas represent fluids. The unshaded areas within the fluids represent "propping rock." The true scale is not known but is shown as though similar to that in Fig. 2.9b. Adapted from Gold and Soter (1984/85).

Figure 2.9b Cross section of what is interpreted to be two vertically adjacent magma chambers (shown in red) beneath Mount Changbai/Paektu, on the border of China and North Korea. The small, nearly empty upper chamber lies at a depth of only 2 to 3 kilometers. Credits to Xu Jiandong, Richard Stone, and *Science* (4 November 2011) vol. 334, p. 587 from which it is adapted. For Mount Changbai/Paektu, also see Fig. 3.27.

Notes and references – *Chapter 2: Deep Carbon and Outgassing*

1 This may change with the advent of technologies able to distinguish among isotopologues. Isotopologues are molecules that differ only in the isotopic composition of their constituent atoms. Light water, H_2O, and heavy water, HDO or D_2O, in which deuterium replaces one or both atoms of hydrogen, are isotopologues of water.

2 Gold [ed. Mitton] (2012) p. 174.

3 N.A. Kudryavtsev [also transliterated Koudryavtsev] (1893–1971), founder of the modern abiogenic theory of the origin of petroleum, noted this by the early 1950s. Subsequent attempts have not changed the situation. According to the conventional biogenic theory, (1) the remains of plankton react with clay to form kerogen and (2) kerogen is then transformed to oil or gas under suitable conditions of pressure and temperature. The second step has been demonstrated but not the first. Versions of the abiogenic theory have kerogen as a residual product from which the light fluids have departed.

4 Ehrlich (2001) Ch. 7.

5 Otto Schmidt's *A Theory of Earth's Origin: Four Lectures* was originally published in 1949 and consists of lectures delivered at the Soviet Union's Academy of Sciences Geophysical Institute in 1948.

6 Mendeleev (1877).

7 Gold (1987, 1999).

8 Ehrlich (2001) p. 134, citing Gold (1993).

9 Kropotkin and Valiaev (1976). T. Gold commissioned an English translation of this paper and published an offer of free copies, for which he received a half dozen requests.

10 Rezit R. Minebaev, Chief Geologist and Deputy General Director of TATEX.

11 Trofimov (2013).

12 Trofimov (2013).

13 Trofimov (2013).

14 Plotnikova, Nosova, and Pronin (in preparation, 2013). The periodic variations of density occurred against a long-term trend of density increase.

15 Plotnikova, Nosova, and Pronin (in preparation, 2013).

16 Plotnikova (2008).

17 Kropotkin and Valiaev (1976).

18 Mendeleev (1877).

19 Ehrenfreund and Cami (2010).

20 A list of locales where carbon-containing molecules have been detected was presented by Steven Soter, an American astrophysicist, at the Kazan workshop on abiotic hydrocarbons, held in Kazan, Russia, 13–17 April 2013.

21 See the *Wall Street Journal* (8 June 1977).

22 So Gold was falsely accused of plagiarizing what was improperly claimed to be wrong-headed science, a real-world version of the joke "The food is inedible and the portions are too small."

23 Hartmann and Davis (April 1975).

24 http://www.psi.edu/epo/moon/moon.html (accessed 27 November 2013). For possible problems with the giant impact hypothesis, see Clery (2013).

25 Robin Canup, personal communication (12 July 2013).

26 Localized indications of heating within carbonaceous chondrites are due to the decay of isotopes that no longer exist, and to collisions in space.

27 Norton (2002).

28 Robert (2001).

29 Condon (1969).

30 Trujillo et al. (2007).

31 See the nonexhaustive list of ice falls in Saul (2006). William R. Corliss's "Sourcebook Project" has additional reports.

32 Endersbee (2005) p. 49. The caption for a map on the same page notes that the petroleum exploration wells in the basin "cover much the same areas as the water bores, indicating a similar primary source."

33 Evans et al. (2013).
34 Guillot and Hattori (2013).
35 Powers and Clapp (1932).
36 Some solar energy reaches such locales indirectly via the constant shower of marine detritus ultimately produced through the mechanism of photosynthesis. Yet, this "marine snow," as it is called, also falls upon the nearby seafloor, where biological activity is tens of thousands of times less than it is at black smokers.
37 Graham (1992). Also see Gould (1994); and Pennisi (2013a).
38 P.N. Kropotkin (1995).
39 Fredrich Wöhler (1800–1882) is regarded as a pioneer in organic chemistry as a result of his accidental synthesis of urea in 1828. "This discovery has become celebrated as a refutation of vitalism, the hypothesis that living things are alive because of some special 'vital force.' However, contemporary accounts do not support that notion. This *Wöhler Myth*, as historian of science Peter J. Ramberg called it, originated from a popular history of chemistry published in 1931, which, 'ignoring all pretense of historical accuracy, turned Wöhler into a crusader who made attempt after attempt to synthesize a natural product that would refute vitalism and lift the veil of ignorance, until 'one afternoon the miracle happened.'" Nevertheless, it was the beginning of the end of one popular vitalist hypothesis ... that 'organic' compounds could only be made by living things." Adapted with minor changes from: http://en.wikipedia.org/wiki/Friedrich_W%C3%B6hler (accessed 12 August 2013).
40 Cited at *origeminorganicadopetroleo.blogspot.com/2011_05_01_archive.html*
41 Petroleum History Society, Canada (Nov. 2002).
42 Coste (1905).
43 Robinson (1963).
44 Dmitrievsky and Valyaev (2013).
45 Dmitrievsky and Valyaev (2013).
46 Dmitrievsky and Valyaev (2013).
47 Plotnikova (2006).
48 Kutcherov and Krayushkin (2010) pp. 20–22.
49 Dimitrov (2003) provides a review of information on mud volcanoes.
50 Kutcherov and Krayushkin (2010) p. 20.
51 "jim z" at, http://answers.yahoo.com/question/index?qid=20110117074920AAvKkA1 (accessed 22 November 2011).
52 Hinrichs et al. (2006).
53 Kutcherov and Krayushkin (2010) p. 6.
54 Alkanes, alkenes, arenes, phenols, aldehydes, carboxylic acids, esters, ketones, nitriles, PAHs, and their halogenated, methylated, and sulfonated derivatives, as well as various heterocyclic compounds including thiophenes and furans, according to Schwandner et al. (2012). Similar results are reported from volcanoes in Kamchatka (Gribble, 2001). Also see Gribble's reply to claims that such gases are "sampling artifacts" (Gribble and others, 1995).
55 Kutcherov and Krayushkin (2010) p. 9.
56 Compare, for example, Gorlenko et al. (2000) with Lukjanova and Lovzova (1994).
57 Kutcherov and Krayushkin (2010) p. 9.
58 The specialized scientific literature on diamonds and their formation is vast. For overviews by nonspecialists, see Kutcherov and Krayushkin (2010), pp. 10–11; and Saul (1981).
59 Kavanagh and Sparks (2009).
60 http://www.innovations-report.com/html/reports/earth_sciences/report-52576.html in which hydrocarbons are translated as "carbohydrates" (dated 6 December 2005; retrieved 24 September 2013).
61 Hitchcock (1865).
62 Hoşgörmez et al. (2008).
63 Kutcherov and Krayushkin (2010) pp. 11–12.
64 Gold (1993).
65 Corliss (Jan. 2010) *Anomaly Register #9*, entry ESC.14.4.

66 Corliss (Jan. 2010) *Anomaly Register #9*, entry ESC.13.36, citing the *Wall Street Journal* (16 April 1999). Also see Trofimov (2013).

67 Trofimov (2013).

68 Ehrlich (2001) pp. 132–134.

69 For comparable occurrences, see Powers and Clapp (1932), Powers (1932), and the sources they cite.

70 Gribble (2001). Also see Gribble's reply to claims that such gases are "sampling artifacts" (Gribble and others, 1995).

71 Corliss (Jan. 2010) *Anomaly Register #9*, entry ESC.14.3.

72 Matthews (17 January 1999), citing Thomas Gold in *The Sunday Telegraph* (London).

73 Corliss (Jan. 2010) *Anomaly Register #9*, entry ESC.14.1, citing W.M. Williams in *Knowledge*, vol. 9, p. 111 (1886).

74 Brothers et al. (2013). Gas escaping from underwater hydrates is for the most part dissolved and does not enter the atmosphere.

75 See Kutcherov et al. (2002).

76 Gold (Dec. 1984); Gold and Soter (Dec. 1979; 1984/85).

77 For a popular presentation of the physics of champagne bubbles, see Ligier-Belair and Polidor (2009).

78 Ehrlich (2001) p. 135.

79 Soter (1999) p. 275. One observation of earthquake lights described them as "gas like and not electrical in nature"; the observation was made in southern Germany 16 November 1911, a time and place where observers would have been familiar with both gas lights and electricity (Schmidt and Mack, 1912–1913).

80 Soter (1999) pp. 280–282.

81 Soter (1999) p. 282.

82 Liu et al. (2009). Paul Silver and colleagues had earlier reported a correlation between barometric pressure and seismic events in California and the northwestern U.S.

83 Wakita et al. (1980).

84 Tributsch (1982).

85 See the appendix to Gold and Soter (1984/85) and the sources they cite, as well as the compilations in the publications of the "Sourcebook Project" of William Corliss.

86 Tributsch (1982).

87 Tributsch (1982) p. 5.

88 Dawson (1970).

89 For materials in the above paragraphs, see Gold and Soter (Dec. 1979; 1984/85); Gold (Dec. 1984, 1987, 1999); and Saul (1981). Diamonds still retain many of their secrets. This highly unusual 22.37 carat specimen from a supposedly Congolese locality is of a gem diamond within a vein-like matrix of non-gem diamond. The "vein" is approximately 7 mm wide at its thickest, and 3 mm at its most thin. It is not clear how such a specimen formed, whether a kimberlite eruption was involved, and if so, how its two components could have reached the surface of the Earth still attached to one another.

In a scenario that is not necessarily in conflict with Gold and Soter, Russell et al. (2012) propose that kimberlites acquire their buoyancy from exsolved CO_2 when a carbonatite-like melt reacts with wall rock, causing a marked drop in its ability to hold CO_2. "The solubility drop manifests itself immediately in a continuous and vigorous exsolution of a fluid phase, thereby reducing magma density, increasing buoyancy, and driving the rapid and accelerating ascent of the increasingly kimberlitic magma."

90 Information is not available concerning the possible existence of still lower magma domains.

3

Early Impact Scars

... as with so many persistent puzzles, the resolution does not lie in more research within an established framework but rather in identifying the framework itself as a flawed view...

—Stephen Jay Gould, *Bully for Brontosaurus* (1991)

During the brief history of meteorite impact geology, all past predictions about the importance of impacts and the range of their effects have turned out to be inaccurate and unimaginative underestimates.

—Bevan French, *Traces of Catastrophe* (1998)

I ONCE WROTE an article that was accepted by the journal *Nature* and then picked up by the daily newspapers.

On reading it, one colleague suggested it was a hoax. Another accused me of having a very active imagination, which, though true, had not been meant as a compliment. And there were three scientists, all previously unknown to me, who intimated that I had stolen their ideas. More happily, I also acquired a dozen enthusiasts. Some were "groupies," a word much in vogue in those days in the late 1970s, but there were also a few senior scientists who might have been unhappy to learn that they had made common cause with groupies.

Mostly, however, there was silence.

Years later, during the coffee break at a scientific meeting, a fellow spotted my name tag and came up to me. "So you're the fellow who wrote about those giant circles. I read the paper when it came out. It was fascinating, absolutely fascinating, wonderful stuff ... *of course it can't be true.*" And he wandered off.

The problem was that I had observed and described something that appeared to go against accepted theory. Theory had then trumped observation. Science is not supposed to work that way but sometimes it does, perhaps even frequently.

The paper in *Nature* had been titled "Circular structures of large scale and great age on the Earth's surface"[1] and the follow-up by the Nature-Times News Service had carried my story under the title "Geology: Strange Circles on Earth." What I claimed was that the Earth still retained faint scars akin to those caused by the enormous impacts that had bashed the Moon. This sounds simple, and indeed it is simple, but — and here is where my troubles arose — the theory of plate tectonics appears to rule out the survival of ancient scars of this nature.

Plate tectonics

The theory of plate tectonics emerged in the 1960s and early 1970s. During those years, investigations of the ocean floor and of the Earth's magnetic field enabled scientists to rework the idea of continental drift, a much-debated hypothesis set out by Alfred Wegener a half century earlier. (Wegener himself had died sometime in 1930 while traversing the Greenland ice cap on a dog sled.)

Plate tectonics holds that the outermost layer of the Earth has been broken into a small number of rigid plates that move in different directions and at different speeds in relation to one another. At their boundaries, plates may collide, pull apart, slide past, or grind against one another at an average speed comparable to that at which a fingernail grows. In places where they collide, one plate may pass under another. It *subducts*, a process evoked earlier (**Fig. 1.40**). Plates are essentially of an "oceanic" or "continental" character and in almost all cases it is an oceanic plate that undergoes the subduction. A spectacular exception involving two continental plates is the slowed, but still ongoing, collision of the Indian plate with continental Asia (**Fig. 1.46**). In the past, some continental material from the Indian plate was subducted under Asia, but more was peeled off in slices (nappes; **Fig. 1.51**) that remain relatively near the surface.

Oceanic crust comes into existence along mid-oceanic ridges during undersea eruptions.[2] It sits on uppermost mantle material and the two are subsequently pushed farther and farther from the ridge by later eruptions. They are gradually cooled from the top down by contact with seawater and as they cool, they become progressively more dense. After some 20 to 40 million years at present-day cooling rates,[3] they become about 1½ percent denser than their underlying material. Gravity may then cause the cold oceanic material to sink — to subduct (**Fig. 1.40**) — *provided* that the local crust is sufficiently weak and the counteracting forces of friction not too great.[4]

On maps, places where subduction occurs commonly appear as arclike curves. These two-dimensional features are actually the tops or surface exposures of sloping three-dimensional zones where the Earth's crust had been locally weak.

Some plates seem to have been in existence from exceedingly early geological times — parts of Amazonia and the Canadian Shield, for example — but others apparently come and go. India and Australia, separated by an oceanic volcanic spreading ridge somewhat more than 50 million years ago, are now, for example, believed to ride on the same plate.

Plate-tectonic theory has provided earth scientists with a wonderfully usable overview of how the Earth works, and as a professional geologist, I am not about to argue that it is fundamentally false. True, it leaves an endless number of details to be worked out, but as with the theory of evolution, plate tectonics also hints that in the end everything may make sense.

Photographs from space

On 4 October 1957, I was an undergraduate student at the Massachusetts Institute of Technology. The next day the corridors of the Department of Geology and Geophysics were all abuzz, but it was not simply because the Soviets had beaten us into space and that Sputnik 1 (**Fig. 3.1**) was orbiting overhead. It was also the prospect that soon there would be photographs from space and that we geologists would at long last be able to see the great terrestrial impact scars akin to those visible on the face of the Moon. The existence of such scars was taken for granted by several of the professors I knew. This seemed evident because the Moon and Earth are neighbors in the vastness of the Solar System, so whatever hit the Moon must have also hit the Earth, a line of reasoning and a conclusion that are still entirely valid. Yet when satellite pictures of the Earth became available just a few years later, the circular patterns were not seen. People gradually ceased looking for them, and in time NASA's Paul D. Lowman Jr. would have to insist that "arguing that the Earth had somehow escaped" the bombardment that had battered the Moon was tantamount to "invoking magic or divine intervention."[5] Still, the scars were not seen, and it was said — and said rather convincingly — that the Earth is the only planet in the Solar System with flowing water on its surface and, "thus," that erosion had erased all traces of such scars. Some doubters remained, plus a few scientists who thought they could discern large circular patterns on Landsat images, though not sufficiently clearly to report their observations in print.[6] I myself was not among the doubters because in those days I was most interested in comparing fossils from opposite sides of the south Atlantic and in understanding what was wrong — or perhaps right — with Wegener's theory of continental drift.

Figure 3.1 Sputnik 1.
http://kolekzioner.net/modules/smartsection/item.php?itemid=222

Doubts persisted until 1971 or thereabouts, when the theory of plate tectonics began to acquire its modern form and to focus the attention of earth scientists. Some then presumed that any terrestrial rocks with Moonlike impact scars would have been subducted, while others, with more promising science to do, lost interest in the search. Thus by around 1971 there was little geological talk of such matters except among a few nonprofessionals, some of them quite cranky, who tended to see meteorite-impact scars everywhere.

As it turns out, neither the scientists, nor the sensible amateurs,[7] nor the cranks had fully understood the situation.

"Accepted" criteria

As of early 2014, approximately 185 terrestrial structures of various ages had been confirmed as bona fide meteorite impact sites.[8] Some are easily visible craters, others are more or less severely degraded by erosion, and still others are buried. Numerous variations and complications exist because not all meteorites have the same composition and because the target rocks at various impact sites had different properties. (Rocks may be massive, stratified, permeable, impermeable, "wet," "dry," porous, fractured, under water, etc.)

Crater turned out to be an overly specific name for these varied impact structures, some of which were only recognized from rock fragments retrieved from drill holes. In consequence, the term *astrobleme* was coined in 1960,[9] a matter I regret because the word now seems ideally suited for another purpose.

Each of the 185 terrestrial sites confirmed and recognized as having been formed by the impact of a meteorite has been accepted on the basis of certain strict and carefully defined criteria. Circularity in itself is given little weight. For "unlike their planetary counterparts, terrestrial impact craters are mostly recognized not by morphology but by characteristic shock metamorphic effects."[10]

A few younger astroblemes, such as the famous Barringer Crater in Arizona and some depressions in the deserts of Australia, are immediately classifiable as impact sites because of the presence of meteorite fragments on the ground nearby. But such sites are exceptional because meteorites — whose component minerals had necessarily been stable in the dry and oxygen-free environment of space — rarely survive for long on the surface of the Earth, where water, warmth, and oxygen reduce them to dust or rust.

In the absence of actual meteorite fragments, suspected impact structures are classified as astroblemes if their rocks or minerals show particular types of deformation *produced only in solid materials*, and at the extremely high pressures associated with great shocks.

Such shock effects include microscopic deformations in quartz and other minerals ("planar deformations"; **Fig. 3.3**), minerals that only crystallize at very high pressures, quartz or the mineral cristobalite with "ballen" structure (**Fig. 3.4**), "toasted quartz" (**Fig. 3.5**), coesite (a mineral with the same chemical composition as quartz but with an internal structure that forms only at higher temperatures), and conical fractures known as "shatter cones" whose formation is not well understood (**Fig. 3.6**).

"Shock metamorphic" features such as these, formed in solid rocks and minerals, have been treated as necessary and sufficient for the authentication of terrestrial impact structures.

Some definitions

A ***mineral*** is a naturally occurring solid substance formed by a geological process and with a definable chemical composition and a specific crystalline structure. Its external form reflects an internal structure composed of atoms organized in an orderly array of repeating units. The mineral zircon, for example, is a silicate of the metal zirconium, and all undistorted, uneroded crystals of zircon are square when viewed down their longest axis (**Fig. 3.2**). Quartz, one of the most common minerals in our planet's crust, is an oxide of silicon. Over 4900 minerals were known by late 2012, most of them rare.

Rocks are aggregates of different minerals mixed together in various proportions, and with no fixed chemical formula. Examples include sandstones and granites. The mineral zircon is a common minor constituent of many granites and granite-like rocks.

Figure 3.2 Zircon, Goiás, Brazil, 2 cm x 2 cm. Photo: Tom Epaminondas.

Natural glasses come from molten materials that cooled too rapidly for crystals to form. Glasses have no particular long-range internal structure but may possess small structured regions that mismatch at their boundaries. Glasses have no fixed chemical composition. Obsidian is both a glass and a rock.

Meteorites are natural objects that have originated in space and have survived their fall to the Earth's surface. Meteorites are composed of minerals but may also include trapped gases and "organic" molecules (Chapter 2). Three major categories of meteorites are recognized, "stones," "irons," and "stony irons," and the existence of short-lived meteorites composed of ice is also probable.

Meteors, also known as shooting stars, are the visible paths of meteorites in the Earth's atmosphere or, more commonly, the visible paths of incandescent meteorite-like fragments in the process of being consumed by atmospheric friction on their way toward the Earth's surface.

Figure 3.3 Planar deformation features (PDFs) in quartz.[11] Photo: Ludovic Ferrière.

Figure 3.4 "Ballen" cristobalite.[12] Photo: Ludovic Ferrière.

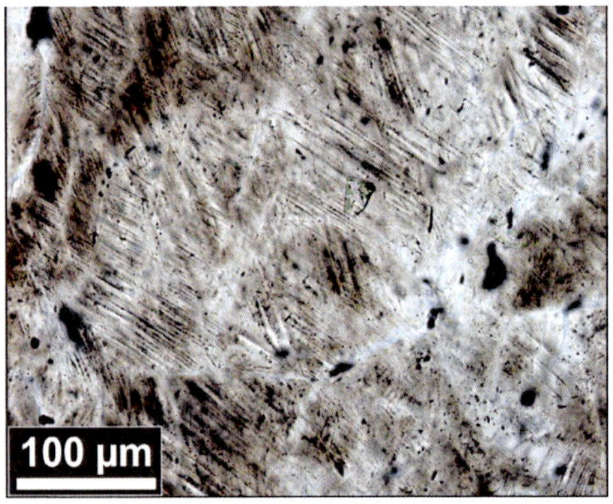

Figure 3.5 "Toasted quartz."[13] Photo: Ludovic Ferrière.

Figure 3.6 Shatter cone.[14] Photo: Ludovic Ferrière.

Table 3.1: Notable meteorite impact craters

Name	Diameter (km)	Age (Ma)
Maniitsoq structure, West Greenland	100(?)	3000± (impact origin awaiting confirmation, Nov. 2013)
Suavjärvi crater lake, Russian Karelia	16	2400 (oldest accepted terrestrial impact crater as of late 2013)
Vredefort Ring, SW of Johannesburg, South Africa	300 (largest recognized)	2023
Sudbury Basin, Ontario, Canada	200	1850
Siljan Ring, NW of Stockholm, Sweden	65–75 (largest in Western Europe)	377
Chicxulub (buried crater), Yucatan, Mexico	170 (multiringed)	65
Barringer Crater, Arizona (formerly "Meteor Crater"), the first impact crater to be recognized as such	1.2	0.05

There were seemingly valid reasons why circular outlines or "craterform morphologies" on their own were not readily accepted as satisfactory evidence of an impact origin. This avoided confusion with the far greater numbers of volcanic craters, and it also served to keep over-enthusiastic amateurs at bay. Yet there is something vaguely illogical here and not quite right. For the same scientists who insisted on being shown fragments of meteorites, melted rock, "planar deformations," or "shatter cones" before giving their imprimatur for a terrestrial impact site were also willing to accept circularity by itself as diagnostic of an impact origin in classifying structures on Mercury, the Moon, and Mars and on the satellites of the outer planets. On Earth, circular structures are "innocent until proved guilty" but elsewhere in the Solar System, they are "guilty unless shown to be otherwise."

Among the oldest of the 185 or so accepted impact craters are the Suavjärvi astrobleme in Russian Karelia, South Africa's Vredefort Ring, and the Sudbury Basin in Ontario. (The impact origin of the still older Maniitsoq structure of West Greenland[15] was awaiting confirmation at the time of writing.) Vredefort and Sudbury are also the largest recognized astroblemes, followed by Chicxulub in Yucatan, formed about the same time as the disappearance of the dinosaurs.

Table 3.1, in which "Ma" designates "million years," a convenient measure of time when dealing with the history of our planet, lists some notable meteorite impact craters. **Table 3.2**, included here for general reference, lists highlights in the history of our planet.

Table 3.2: Selected chronological landmarks and their approximate dates

Events	Dates (Ma)
Age of the Earth.	4540
Formation of the Moon.	4527
Oldest dated mineral grains (abraded zircons from Western Australia).	4404±8
The Late Heavy Bombardment (LHB), a unique episode that scarred the inner planets; the age of the LHB was originally obtained by laboratory determinations on shocked Moon rocks. Indications that the LHB may have tailed off into later times, or included impacting bodies that took long routes before reaching the Earth, come from dated sediments that contain sand-sized remains of droplets condensed from fiery impact plumes.[16]	4100–3800, but perhaps including events 3500–1700
Radiometric ages obtained from parts of the Acasta Formation in northern Canada, our planet's oldest known rocks.	4031±3, 3960, 3580
Oldest of several rock formations with similar dates.	3600+
Oldest astrobleme (**Table 3.1**).	3000± or 2400
Great Oxidation Event[17] or First Great Oxidation Event when the atmosphere first began to accumulate small quantities of oxygen, the waste product of primitively photosynthesizing marine bacteria. In earlier times, such oxygen had been consumed by the ocean's primordial store of dissolved iron, but toward 2400 Ma, bacteria had "rusted the ocean," leaving no iron to capture newly produced oxygen.	probably 2450–2320
Oldest known remains of possibly multicellular life; found in Gabon.[18] Called a "biota" because it is unclear whether these fossil-like structures, if they are indeed the remains of life,[19] are unicellular or, if multicellular, if they are animals, plants, fungi, or perhaps an eerie "none of the above" (**Figure 3.7**).  **Figure 3.7** External and internal views of one of a large number of similar unidentified fossil-like objects from Gabon. Lengths of individual specimens range from 1 to 12 cm. Photo: A. El Albani.	2100±30
Sponge-like fossils from Namibia.[20]	760
Second Great Oxidation Event involving oxygenation of the deep ocean, along with broadly concurrent worldwide erosion, commencing around 550 Ma ago.[21]	635± –510
(i) Sponge-like fragments and (ii) tiny balled-up multi-celled marine or lake-bed fossils,[22] whose initial identification as animal embryos possessing bilateral symmetry has been severely questioned.[23]	635–580 (or 635–551)
Sponge-like fragments from South Australia.[24]	600–580
Abundant worldwide occurrence of the Ediacara biota, which may or may not be the ancestors of living creatures (**Fig. 3.8**). The Ediacarans differ greatly from the 2100 Ma fossil-like objects from Gabon (**Fig. 3.7**), and may be animals, plants, or fungi, unicellular or multicellular. Alternatively, they may be "none of the above," or they may be a failed experiment in multicellular life, as proposed by Adolf Seilacher.[25]	576–542
The Cambrian Explosion, during which our own multicelled ancestors first appeared. All living animals with specialized cells descend from creatures that emerged during this complex multistep episode.	542, with the main phase 530–520
End of the age of dinosaurs.	65
Appearance of the first members of the genus *Homo*.	2.4±0.2
Emergence in Mesopotamia of writing, cities, kingship, and much else that characterizes "High Culture."	0.005
Emergence of modern nation states in the Western world.	0.0002

Figure 3.8 Images of representatives of the Ediacaran biota showing different types of symmetry or apparent symmetry. Although the Ediacarans are generally held to lack left-right symmetry, the fossils shown in (f) and (i) suggest that this may not be strictly true. Length of scale bars: 1 cm. Details for each image are given in a note.[26] Credit: Shuhai Xiao (Virginia Tech) and Marc Laflamme (University of Toronto Mississauga).

The Late Heavy Bombardment

The Late Heavy Bombardment (LHB), referred to in the previous chapters and in **Table 3.2**, is the name of the episode that battered Mercury, Venus, Mars, the Moon, and necessarily the Earth as well. Evidence used for dating the LHB has come more or less exclusively from determinations made on rocks that were melted by impacts on the Moon or ejected from lunar impact craters. (When sufficiently heated, as by impact-shock, rocks liberate certain accumulated products of radioactive decay, thereby resetting their radiometric clocks at zero.) Almost all of these lunar rocks were shocked or melted during a single interval between 4100 and 3800 million years ago, which is to say somewhat before the consolidation of the oldest known rock formations here on Earth, several of which are dated around 3600+ Ma (**Table 3.2**). This said, parts of the Acasta Formation of northern Canada are slightly older (**Table 3.2**), suggesting that they are survivors of part or all of the bombardment.

The Earth underwent the same battering as the Moon, its neighbor in space, much as a rainstorm soaks both members of a couple standing together in an open field. Yet before the dating of the shocked Moon rocks, geologists had imagined that the scarring of the Moon had been the result of occasional meteorite falls that had intermittently occurred throughout the history of the Moon. No one had envisaged a single long-ago long-extinct bombardment episode.

There was a historical reason that the entire geological community* failed to anticipate such a bombardment. For James Hutton (1726–1797), one of the most influential founders of the science of geology, had famously claimed "that we find no vestige of a beginning — no prospect of an end."[27] In consequence, geologists believed that there should have been no bombardment at the beginning because there had been no beginning. Yet Hutton had been misunderstood. He had not meant that there had been no ultimate beginning. He had simply meant that a lot has happened in the course of the "deep time" in which geologists bathe and that no ultimate origin can be detected through the many cycles of volcanism, sedimentation, and erosion.

Although Hutton had been nearly universally misread or misunderstood, his concept of deep time and his "no vestige of a beginning" have nevertheless helped generations of geologists to get a feeling for their subject. As it turns out with the discovery and dating of the Late Heavy Bombardment, we actually do have an intelligible beginning to our planet's history, starting around 3800 million years ago.

* Or, more accurately, by the entire English-speaking geological community, plus many others.

The slightly older dates from the Acasta Formation do not seem to contradict this conclusion, but some considerably older grains of zircon with radiometric ages of around 4400 million years have also been discovered (**Table 3.2**). These require a separate discussion, but it otherwise appears that cosmology stopped and legible geology (the geological record) commenced some 3800 million years ago or somewhat later, toward 3600+ million years ago, after our planet had cooled down following the bombardment.

The Nice Model

Once upon a time there was a Greek deity named Νίκη who was the goddess of victory. Her Roman equivalent was Victoria. Νίκη does not seem to have done very much, not even in myth, but her memory lingers in the names of Nike sports equipment and of the French Mediterranean city of Nice.

It was at the Nice Observatory beginning around 2005 that a group of scientists joined forces and devised a seemingly realistic explanation for the Late Heavy Bombardment.[28] The "Nice Model," as it is called, is complicated and is not fully accepted, but it has nicely and victoriously withstood several volleys of criticism. Also in its favor is that it originally had little competition; older, simpler explanations for the LHB had failed to make much sense.

Briefly, the Nice Model says that the familiar configuration of the Solar System (**Fig. 3.9**) is not the same as it was before the LHB. In those times, the giant planets Jupiter and Saturn moved around the Sun in orbits quite different from those we now observe. In addition, an enormous disk composed of chunks of rock and ice, with a mass estimated to have been about 35 times that of the Earth, then orbited the Sun in the outer regions of the Solar System beyond Jupiter, Saturn, Uranus, and Neptune, and a very long way from the Earth and Moon. The inner portion of this rubble disk would have been located outside the orbits of the giant planets Jupiter and Saturn.

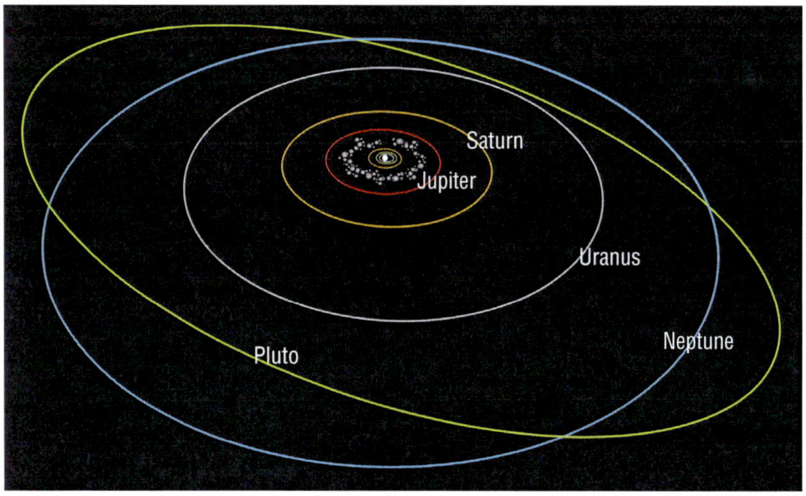

Figure 3.9 The Solar System with the present-day orbits of the inner and outer planets shown to approximate scale. The orbits of Mercury, Venus, Earth (shown in green), and Mars are too small to be labeled. The cloudy effect inside the orbit of Jupiter represents the asteroid belt.
©Chris Impey / http://m.teachastronomy.com/astropediaimages/LayoutSolarSystem_all.jpg

The force of gravity decreases in proportion to the square of the distance between objects. In consequence, the inner portion of the rubble disk interacted far more forcefully with the two giant planets than did the outer debris. As a result, orbiting chunks from the inner parts of the disk moved inward while the orbits of the giant planets migrated slowly outward.

Saturn is only 30 percent as massive as Jupiter. Hence it migrated a much greater distance than Jupiter during the hundreds of millions of years between the formation of the Solar System and the commencement of the LHB (**Table 3.2**). Eventually the orbits of these two giant planets came into a "2:1 resonance," a configuration that caused Jupiter plus Saturn to give a combined gravitational tug at each orbital passage, thereby disrupting the orbits of Uranus and Neptune.

Uranus and Neptune then scattered material from the outer portions of the debris disk, with a series of planetary resonances sweeping through the debris belt, sending rock and ice in all directions and heavily bombarding the inner members of the Solar System. Some calculations indicate that the episode scattered more than 99 percent of the primordial debris. This was a unique, one-time, never-to-be-repeated set of cosmological circumstances. See **Figure 3.10**.

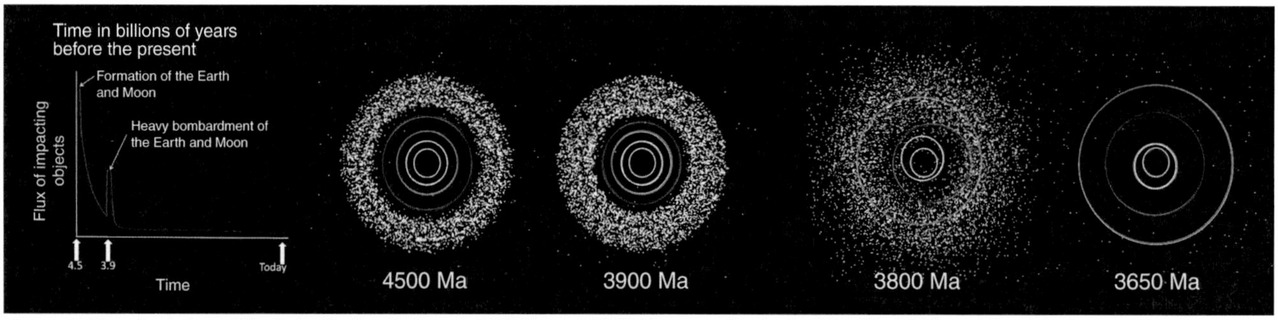

Figure 3.10 Evolution of the Solar System between 4500 Ma and 3650 Ma according to the Nice Model. The orbits of Jupiter and Saturn are shown in white, with Jupiter innermost. Neptune's orbit is shown in blue, Uranus' orbit is in red. The orbits of the inner planets are not shown. Adapted from "The Impact History of Asteroidal Meteorites" by Rhiannon Mayne, *Meteorite*, vol.19, no.3 (Fall 2013) pp. 26–28.

The LHB was an episode, not an event. A multitude of discrete events occurred with some things happening before others. For the time being, many details elude us. In some scenarios, the most massive impacting bodies reach the region of the Earth and Moon toward the very end of the LHB. In others, the LHB lasts for about half of our planet's history, tailing off with a few late arrivals toward 3500 to 1700 million years ago.[29]

The Nice Model is not a "Theory of the Late Heavy Bombardment" as such. It is a far more comprehensive concept that also attempts to explain the absence of much material outside the orbit of Neptune, the present-day positions of the outer planets, the tilts of their orbits, the degree of noncircularity (eccentricity) of their orbits, the capture of certain moons, and much more. Ongoing tests of the Nice Model involve computer runs that simulate hundreds of millions of years of orbital interaction, each one with a slight perturbation from the previous. Computer runs are abandoned, reconceived, or reprogrammed when the results produce unrealistic outcomes, as when asteroids wind up with orbits tilted at angles greater than those actually observed. The Nice Model may well be valid, but a final version is not likely to be ready for some time.

As noted in the previous chapter, the planets nearer the Sun were formed at the high temperatures at which iron, silicon, aluminum, and other more-or-less refractory (nonvolatile, high-melting-point) elements condense. In consequence, these planets, Mercury, Venus, Earth, and Mars, are for the most part rocky. By contrast, bodies in the outer regions of the Solar System, formed in far colder circumstances, are chiefly composed of gases and ices that do not condense unless the temperature is very low. (Rock "freezes" before ice.)

Among its numerous predictions, the Nice Model indicates that the LHB had flung solid chunks of low-density materials such as methane and water into the inner regions of the Solar System. At first it was not evident whether much, or any, of this ice had survived to furnish our rocky planet with water. But, as cited in the previous chapter, measurements on meteorites whose compositions represent the make-up of the inner Solar System subsequently resolved the matter, showing that the ratio of naturally heavy water to normal water here on Earth is markedly different from that of the water formed in the inner Solar System. "The terrestrial water must therefore have been imported from the coldest regions of the solar system after the Earth had formed."[30]

"The most common landform in the Solar System"

Impact craters are extremely common throughout the inner Solar System. Some 5185 lunar craters with diameters of at least 20 kilometers have been mapped[31] and one count of Martian crater scars with diameters greater than one kilometer came to 635,000.[32] The Earth, has been calculated to have had 2500 to 3000 impact scars greater than 100 kilometers in diameter.[33]

Matters are similar on almost all members of the Solar System possessing solid (i.e., rocky as opposed to gaseous) surfaces. Differences between one rocky planet or moon and another are due to local conditions: the strength of the planet's or moon's gravitational field, the nature of its atmosphere and crust, history of volcanism, and distance from the Sun, and possible additional considerations that may come to light as the Nice Model is refined or replaced. An encounter with an exceptionally large impacting body would have had a particularly great role.

The only nongaseous planetary body lacking visible impact scars is Jupiter's innermost large moon, Io. Io's proximity to Jupiter subjects it to great tidal forces with each rotation, "Io tides." Consequent stretching and relaxing generates frictional heat, producing molten rock that repaves Io on a nearly constant basis and has apparently eliminated all nonvolcanic features.

The great majority of larger impact scars throughout the inner Solar System were formed during the Late Heavy Bombardment. In the case of the Moon, the LHB is thought responsible for 75 to 90 percent or more of the now-visible craters with diameters greater than 20 kilometers.[34] This, however, does not properly reflect the severity of the LHB, an episode that lasted hundreds of millions of years during which an unknown number of earlier-formed craters were obliterated by later impacts. "Impact saturation" is said to occur when each new impact destroys an "equal amount" of previous impact scarring, a matter easy to understand for a two-dimensional planetary surface but probably impossible to define if the depth dimension is taken into consideration.

Around 3850 million years ago, the LHB cratering rate on the Moon — and necessarily on the Earth too — was an estimated 1000 to 1500 times the rate during the period that produced the 185 accepted terrestrial astroblemes.[35] The Earth's surface would have then been saturated by impact scars, "the most common landform in the Solar System."[36] Yet the 185 currently accepted impact sites were without exception formed long after the Late Heavy Bombardment; Suavjärvi, the oldest presently recognized meteorite impact site (**Table 3.1**), dates to one and a half billion years after the end of the LHB.

The Earth has been affected by two different groups, two very different statistical populations of impacts, *a young population* represented by the 185 recognized astroblemes of the last 2400 Ma, and *a very much larger old LHB population* that affected the entire Earth to some unknown depth during the period from 4100 to 3800 Ma. It is unclear to what extent knowledge of the young population, which has been intensely studied, might be of use in understanding the far older and far more violent Heavy Bombardment.[37]

The end of Late Heavy Bombardment provides a reasoned starting point for geology, a Time Zero when cosmology and planet-building can be said to have given way to geology.

Yet matters are not entirely straightforward. First of all, more-or-less familiar geological processes were necessarily going on during the course of the LHB. In addition, there are known survivors of the bombardment, parts of the Acasta Formation somewhat older than 4000 Ma, and grains of zircon (a highly resistant and refractory mineral) from Australia with ages of crystallization up to 4404 Ma (**Table 3.2**). Other survivors of the bombardment may exist among mineral grains whose ages cannot be radiometrically established.

The ancient zircons

Many recently fallen meteorites have been radiometrically dated to determine how long ago they cooled to the point where certain products of radioactive decay could not escape. The very oldest dates fall into a cluster and, given that all known meteorites are members of the Solar System,[38] the existence of this cluster has allowed specialists to calculate the age of the Earth at about 4540 million years, almost a half billion years before the commencement of the LHB.

A uniquely violent event occurred toward the beginning of the half billion years preceding the LHB (which was itself also unique). A near or glancing collision with a Mars-sized object[39] caused our planet to acquire its moon around 4527 million years ago. Opinions vary, but the birth of the Moon most probably melted the Earth's primordial crust down to an exceedingly great depth. Yet by sometime well before 4400 Ma, a second-generation crust had developed. We know this because the 4404 million-year-old zircons from

Australia are typical of zircons found in granites, and granite, as discussed later, is a type of rock that solidifies at depth within the Earth, not by the outpouring of lava.

The existence of these zircons indicates that by around 4400 million years ago the Earth possessed a solid crust with a significant but uncertain thickness. Impacts on such a surface would have left scars.

Special maps

Geologists (who spend much of their time making maps) and the specialists who provide satellite imagery are tasked with providing their employers and the public with as much information as possible. But as it turns out, the well-filled maps they produce are commonly ill suited for viewing vestiges of the LHB.

When my paper in *Nature* came out in 1978, one enthusiastic believer from within the mainstream scientific community was William H. ("Bill") Byler, a retired engineer. After several airmailed exchanges and one big package of maps that arrived by sea, Bill confessed that he was having better luck spotting circular patterns on "junky sources" — his phrase — such as children's atlases, airline route maps, and advertisements, and even on black-and-white photocopies of geological maps originally produced in color. In time, I came to recognize the truth of his observation and eventually recalled a similar example from years in Africa. In those times I occasionally listened to the Voice of America in *S P E C I A L E N G L I S H*, in which the world news was read slowly, with a limited vocabulary, and with pauses between words. It seems that everyday colloquial speech is too cluttered for certain purposes and this also turns out to be the case with detailed geological maps and sophisticated satellite imagery in which the immensely detailed contents may conceal an underlying essence. I also think of the Bollywood films where the unalerted viewer may not grasp that the long minutes of song, dance, and flower petals mask the simple fact that the boy has propositioned the girl.

Byler was not alone. In 1961 the now-obvious rings around the Moon's Orientale impact basin were detected by W.K. Hartmann and G.P. Kuiper only after Kuiper had set up a globe and a projection system, allowing the two of them to "back away" and finally see "gross features … that are missed in the usual quest for more detail."[40]

Figures 3.11 to **3.13** show circular structures observed on "junky sources."[41]

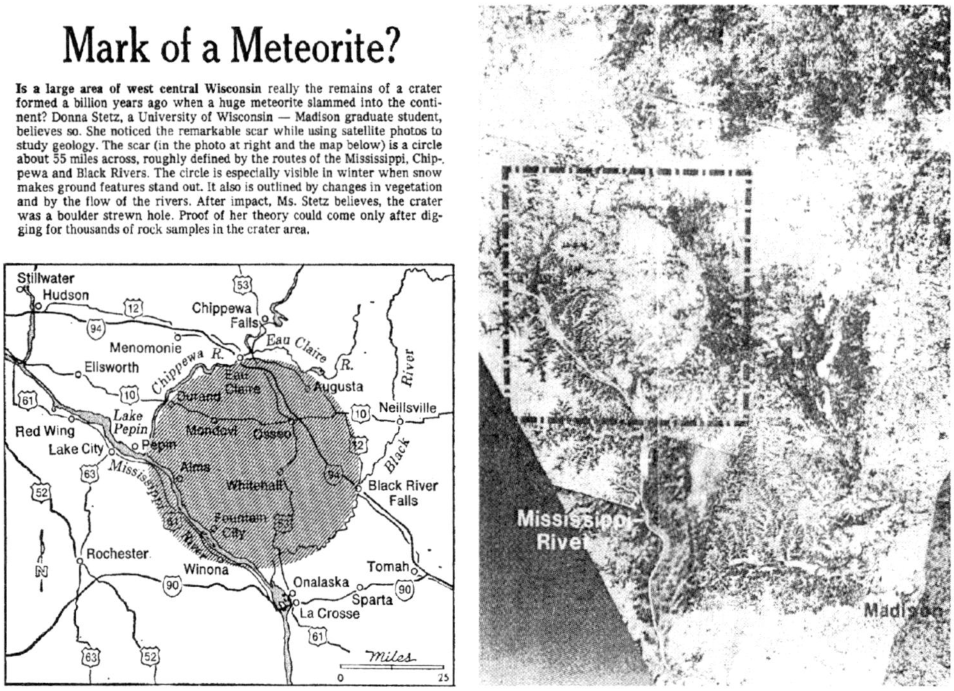

Mark of a Meteorite?

Is a large area of west central Wisconsin really the remains of a crater formed a billion years ago when a huge meteorite slammed into the continent? Donna Stetz, a University of Wisconsin — Madison graduate student, believes so. She noticed the remarkable scar while using satellite photos to study geology. The scar (in the photo at right and the map below) is a circle about 55 miles across, roughly defined by the routes of the Mississippi, Chippewa and Black Rivers. The circle is especially visible in winter when snow makes ground features stand out. It also is outlined by changes in vegetation and by the flow of the rivers. After impact, Ms. Stetz believes, the crater was a boulder strewn hole. Proof of her theory could come only after digging for thousands of rock samples in the crater area.

Figure 3.11 Snow cover, Wisconsin; diameter of the circle, approximately 88 km. From an unknown newspaper, circa November 1977, with thanks to Donna Stetz[42] and Betty Goldman.

Figure 3.12 Ordos platform, Inner Mongolia, north China. Source: Li Jianhua and Shen Xuhui (2001): "Syntaxis faults at the northeastern corner of the Tibetan Plateau and the annular tectonics in southern Ordos," *Journal of Seismology and Geology*, vol. 23(1); http://d.wanfangdata.com.cn/Periodical_dzdz200101015.aspx (3 December 2010), with thanks to Hope Zheng Yan.

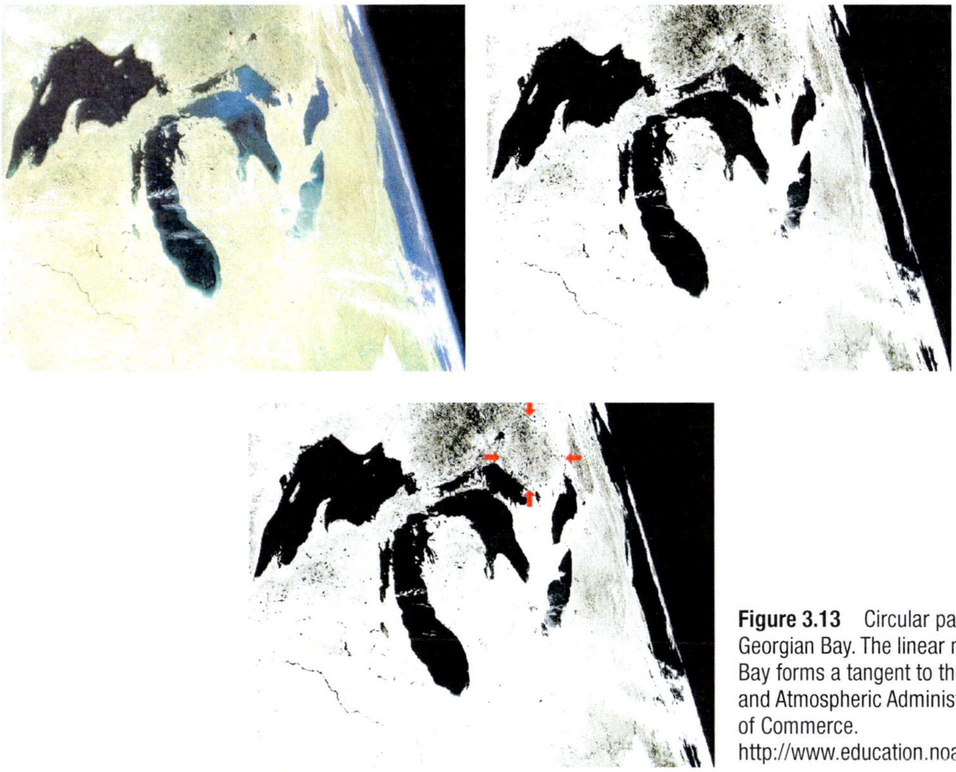

Figure 3.13 Circular pattern north of Lake Ontario and Georgian Bay. The linear northeastern shoreline of Georgian Bay forms a tangent to the circle. Source: National Oceanic and Atmospheric Administration (NOAA) and the Department of Commerce.
http://www.education.noaa.gov/images/greatlkecoregion.jpg

For large circular patterns, globes may be more useful than maps. One of my favorites is a monochrome globe with 3-D relief designed to teach geography to blind students.

Initial conditions

The LHB established "initial conditions" for our planet, and these conditions directed and constrained later geological processes. The circular fracture-patterns — *many of which are difficult to see and easy to ignore* — are prime features of the original terrestrial canvas, a battered, punctured, and much patched and repaired canvas on which all later geology has been painted and overpainted.

As discovered or rediscovered in the early years of the computer age, complex systems may be highly sensitive to their initial conditions. Examples of initial conditions whose effects may linger forever after include the health of a newborn child, the natural resources of a newly populated island, the preparation of an army going to war, and the correct or incorrect demarcation of a property line or an international border. In the particular case of the Solar System, initial "...events like giant impacts early in solar system history, can send planets down vastly different evolutionary paths."[43]

Simple craters and complex craters

Among the 185 accepted terrestrial impact craters (as of early 2014), those with diameters less than 2 kilometers, and some that are a bit larger, are bowl-like "simple craters." Larger impact craters are termed "complex" and are characterized by central peaks and, in some cases, by one or more outer rings. The transition from simple craters to complex occurs at different diameters for each planetary body, determined primarily by the strength of the gravitational field but also locally by the type of target rock and other factors. The surviving terrestrial LHB scars discussed in this chapter are all from impact craters well over 5 kilometers in diameter and are all presumed to be complex.

Violently rapid expansion of hot gases and other phenomena involved in the formation of impact craters are much the same as those involved in explosions, which is to say, situations in which the materials involved cannot get rid of energy fast enough.* Impact craters are in fact a type of explosion-crater, and in consequence, virtually all impact craters, simple or complex, are circular. The few exceptions, not further discussed here, are caused by impacting bodies that arrived at exceedingly low angles, less than 5 degrees above the horizon.[44]

The secondary rims around complex craters at first look somewhat like what one sees when a rock is thrown into a pond. In the past, some thought that outer rings formed when impacted rocks underwent liquefaction and then abruptly froze in place, producing frozen ripple marks. Subsequent observations showed that in detail the concentric rings and scarps on the Moon are irregular and ragged, and do not really resemble frozen waves. They had actually been formed by the inward flow of partly liquefied material immediately after impact, followed by fracturing and inward slumping of the overlying brittle crust.[45] Such structures have also been recognized on Jupiter's moon Callisto[46] and on Mars, where the "concentric rings are not simply uplifted material but correspond to long-lasting and deep-seated breaks in the crust, faults that become progressively less well defined away from the basin center."[47]

The patterns are not geometrically neat or complete, one specialist referring to "the embarrassing fact that the original rim of very large [lunar] basins such as Orientale cannot be found (or agreed upon)."[48] It appears that the inward slumping of one portion of a ring may interfere with adjacent sections, preventing them from slumping, moments later, in quite the same geometrical manner.

Transient craters

The maximum depth of the disturbance produced by a high-energy meteorite impact is not the depth usually seen, measured, or illustrated. Instead, it is the considerably greater depth of the ephemeral "transient crater" produced at the instant of impact, and most readily discerned in laboratory experiments and computer calculations.

Following impact, the bottom of the crater rebounds. A 3-D boundary then comes into existence between (1) those materials that had undergone liquefaction and rebound, and (2) the less-affected rocks that had only undergone fracturing. This boundary corresponds to the transient crater, and its shape and dimensions are of far greater interest than those of the observable post-rebound craters because the fragmented rocks (but not those that had

* Irregularly shaped holes, punched in rooftops or the ground by small meteorites, do not involve explosions and are not classified as craters.

undergone liquefaction) subsequently serve as 3-D pathways for fluids of various sorts.

Observations of simple craters show ratios of depth to diameter of about 1 to 3. Contrary to what might be expected, complex craters are far shallower, a consequence of their far greater production of melted materials, and their depth-to-diameter ratios are roughly 1 to 10. **Figure 3.14** shows a profile of a complex crater. The thick line indicates the transient crater at the moment of impact, and the thin line shows the much shallower resultant crater at a later time.

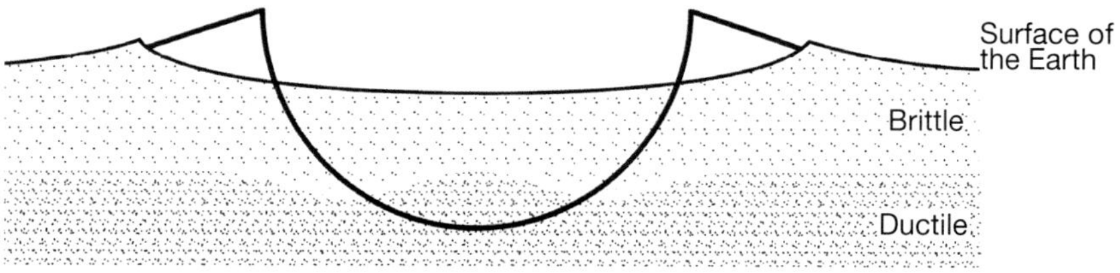

Figure 3.14 Highly simplified profile of a complex crater. The thick line indicates the initial hemispheric excavation, and the thin line shows the much shallower resultant crater at a later time. Vestiges of a central uplift are schematically indicated but outer rims are not. No scale.

Transients of various sorts, in both time and space, are well known to engineers. A familiar transient is the interval between the flip of the switch and the time an electrical appliance begins to function.

Other instances of liquefaction

In rough seas, a ship's cargo of granular solids with air between the grains can briefly, i.e., transiently, acquire the qualities of a fluid. The cargo can then flow to one side of a wave-tossed ship, where it may instantaneously regain the properties of a solid and not flow back. The ship may then capsize. Similar processes take place in mixtures of sand and groundwater during earthquakes, leaving traces that allow some ancient earthquakes to be dated. Intentional liquefaction and fluidization have numerous applications in manufacturing and heavy industry.

LHB scars associated with the North American, South American, and European cratons

When a large crater-forming meteorite strikes, it produces molten material that is partly derived from the impacting body and partly from the target. As might be expected, larger impacting bodies and those arriving with greater velocities produce *proportionately* greater amounts of molten material. Yet calculations show that for sufficiently great impacts, the volume of melt rock is *disproportionate*. In consequence, the entire excavation may be filled[49] *past the point of overflowing*. H. Jay Melosh, who has pioneered such calculations, has shown that for the Earth, an impact excavation might overflow if its transient crater had a diameter of about a thousand or somewhat more kilometers.[50]

It would be easy to suppose or to assume that a 3800–4100 million-year-old shock feature a thousand or two thousand or more kilometers across would retain rocks containing some of the accepted indications of shock metamorphism such as those shown in **Figures 3.3–3.6**. It would also be easy to suppose that all such evidence had been obliterated in the course of geological time. As it turns out, neither assumption is safe. For indications of shock metamorphism could not have been retained, nor could they have been erased, if they had never existed. Instead, as appears entirely possible, the enormous and *disproportionate* amount of melting at such sites would have greatly limited the amount of solid rock available to undergo shock metamorphism.[51] In such circumstances, features such as those in **Figures 3.3–3.6**, which form in solid materials (and not in fluids), and are used for identifying shocked rocks, may have never existed in abundance, or perhaps not at all. This is an important point: *at very large impact sites, the prime effect would have been the production of melt. In consequence, the accepted diagnostic shock features, which form only in solid rock, may not have been produced in abundance.* Or if they had ever existed in abundance, it may have been only as transient features, immediately melted or annealed.

Very large impacts would have produced melted rock that completely filled the rebounded crater, *greatly overflowing its edges*.

Immediately following the bombardment, and for a long time after, the (1) *molten crater fill* and the (2) *melt overflow* from these very large sites would have appeared much the same if viewed from space. The circular boundary of the crater itself would not become evident until the thickness of the solidified melts had been reduced by erosion, and movements from deep in the Earth had regenerated the fractures that delimited the transient crater, piercing the overlying rocks.

Molten materials within the very largest LHB impact sites remained fluid far longer than at smaller impact sites, a consequence of their greater ratio of volume to surface area, another toast vs. crumbs argument. Later-arriving LHB impacts, if not overly large, might have *physically* disappeared into the giant long-lived melt-puddles. But the Nice Model indicates that LHB impactors had formed at various distances from the Sun. In consequence, they would have had varied chemical compositions. They would have thus generated local variations in the *chemistry* of the melts into which they fell.

When a giant impact site and its overflow regions finally solidified, it was from the top down, forming a great expanse of essentially unfractured rock of somewhat spotty chemical composition. Concealed within these rocks was a swamped and buried 3-D craterform whose 2-D circular outline would not become visible at the Earth's surface for a long time after.

As the Earth continued to cool, losing heat into space, the melts solidified, forming various types of basalts, a family of relatively dense volcanic rocks that typically weigh 3.0 tonnes per cubic meter when cool, but less when hot. Vast, deep expanses of unfractured warm basaltic rocks came into being.*

Then, as the basalts continued to cool — becoming more dense in the process — they trapped hot underlying materials, insulating them from the cold of space. But the general cooling of our planet's hot interior and the decay of radioactive isotopes (more important in the past than now) would in time produce more heat than could be dissipated through the insulation of the cold basaltic cap. Reheating from below would have ensued, leading to *remelting at depth of some of the already solidified basalt.*

Rocks — basalts in this case — are mixtures of different minerals, each with its own melting point, so the initial remelting of the basalts was only partial, commencing with those minerals with the lowest melting temperature.† Droplets of the most readily melted minerals formed, relatively rich in silicon, water, and other low-density materials, hence less dense than the original basalts. With time, the drops accumulated, coalesced, and rose by buoyancy. Solidifying, they formed scattered island-like units of rocks of the granite family, a type of rock that typically weighs 2.6 to 2.7 tonnes per cubic meter.

I envisage the gradual transformation of a three-layered structure into two layers. At the bottom, there originally was a dense melt, upon which there floated a relatively light granitic melt rich in water and silicon. These two melted layers were capped by solid basalt, a rock that was heavier than the underlying granitic melt, and whose upper regions were progressively becoming more dense as they continued to cool. Such a situation is gravitationally unstable, and as the lighter granitic fluid rose, it pushed up and fractured mountains of basalt. The basalt mountains then foundered, leaving a stable two-layered configuration with granitic mini-continents floating on heavier materials below.

The mini-continents, built by chemical reactions from the bottom up, were very deep-rooted. Thus, although their chemical compositions at depth were indeed broadly granitic (or granodioritic with typical densities of 2.8 to 2.9 tonnes per cubic meter), the actual rocks they formed would have had a sort of banded structure, a consequence of pressure and folding. Such rocks are termed "granitic gneiss" rather than "granite";[52] (**Fig. 3.15**), and the Acasta Formation (**Table 3.2**) is commonly referred to as the Acasta Gneiss.

The quantity of granitic rock produced by this type of process amounts to only 5 to 10 percent of the initial mass of basalt. The remaining 90 to 95 percent of the initial basalt, having lost its light constituents and thus even more dense than before, would sink into the Earth's mantle to be replaced by fresh basalts still in possession of their light constituents. Eventually, a patchwork of broadly similar deep-rooted granitic and granitic-gneissic units *of various ages* would be created.

* The formation of these basalts would have generally been a two-step process, dense magnesium- and iron-rich rocks of the peridotite family forming first, sinking under their own weight, partially melting, and eventually giving rise to fluids with the composition of basalts.

† Chemical variation within minerals also plays a role in determining the melting point: fayalite, the pure iron variety of the mineral olivine, $(Mg,Fe)_2SiO_4$, melts at around 1600°C at atmospheric pressure, whereas forsterite, the pure magnesium variety, melts at around 1900°C. Gem olivine, which is nearly pure forsterite, is called peridot in the gemstone trade.

Figure 3.15 Polished slab of rock showing granite (upper left) in contact with granitic gneiss, which it has partly invaded. The red spots are non-gem garnets. (Shown actual size.)

Dense materials sank and light materials rose, and in time a 3-D body corresponding to the original circular scar, however modified around its overflow edges, would have come into existence. The result would be a continent-sized body composed of irregularly shaped deep-seated granitic units, each with its own age of solidification and particular chemical characteristics. Areas between the granitic units might show indications of shearing, thrusting, suturing, sequential shallow subduction,* thermal modification, and chemical interchange. This is what we observe today, eroded down to a depth where pressure had firmly welded each unit — or "terrane" — to its neighbors and where granitic gneisses predominated over granites.†

In this manner, each sufficiently large crater was transformed into a mosaic of low-density, quartz-rich granitic (gneissic) rocks within which deep faults and fractures were generally absent. These mosaics of terranes, each bounded by a great 3-D craterform fracture pattern, became the stable cores of the continents.

Remnants of these patterns can still be discerned. They are called cratons, and are conventionally understood as old and stable parts "of the continental lithosphere. Having often survived cycles of merging and rifting of continents, cratons are generally found in the interiors of tectonic plates. They are characteristically composed of ancient crystalline basement rock [granitic gneisses and others], which may be covered by younger sedimentary rock. They have a thick crust and deep lithospheric roots that extend as much as a few hundred kilometers into the Earth's mantle. The term craton is used to distinguish the stable portion of the continental crust from regions that are more geologically active and unstable."[53] Cratons are composed of "shields" in which the basement rocks crop out at the surface and "platforms" in which the basement is overlain by younger rocks and sediments. Cratons are far less heavily fractured than other parts of the Earth's continental crust.

Cratonic shields may also have a particularly ancient core, termed an "archon." Some archons, not all, are situated near the centers of their respective cratons, but the surrounding provinces do not give the appearance of having been added or accreted concentrically in the manner of a hailstone or stalagmite. This is consistent with their mode of formation proposed here. Granitic units — whether called terranes or provinces — were individually derived from basaltic units 10 or 20 times their size. They were subsequently eroded to a level where pressure and temperature had transformed granite to granitic gneiss. At

* Subduction on the early hot Earth, and for a long time afterward, appears to have been characterized by shallow events in which the ends of downgoing slabs broke off. If so, this might have been due to a lesser difference in temperature (and hence density) between the descending unit and overriding unit, both of which would have been hotter than in present times.

† Lack of sufficient internal heat hindered the development of lunar granites.

these pressures and temperatures, individual units of granitic gneiss became welded to one another.

The Canadian Shield, situated within the North American craton, has been intensely studied, a large task made easier by the fact that much of the bedrock has been scoured by glaciation. In his innovative review paper "United Plates of America, the Birth of a Craton," Paul F. Hoffman discerned some 15 ancient provinces welded together, each with its characteristic radiometric age,[54] ranging from perhaps 3580+ million years to about 1250 million years (**Table 3.2**). These terranes are old, but the impact scar that produced the North American craton is older still. *The scar patterns are older than the rocks used to define the cratons,* much as a childhood scar is forever older than the skin cells of the adult. Impacts that occurred 4100 to 3800 million years ago produced scars that are these days manifested by rejuvenated fractures and by rocks that are only 3580 to 1250 million years old (as well as by still younger sediments). The fractures are older than the rocks in which they are observed.

Cratons are "very robust entities indeed."[55] North American, South American, and European craton scars are shown in **Figures 3.16, 3.17,** and **3.18**.

Note: Some of the maps and other imagery used in the remainder of this chapter are based on non-conformal map projections or otherwise less-than-ideal maps. This has necessarily introduced some distortions and geographical inaccuracies, especially when using maps that cover large areas, but it does not negate my overall argument. Such distortions and inaccuracies may be minimized in future research by producing maps and globes designed especially for the purpose. I have not done so because it would counter my contention that many of the scars left by the Late Heavy Bombardment can be seen with no need for specialized techniques.

The circular patterns shown in the remainder of this chapter come from a broad variety of sources, scientific and not. Those who reject the idea that vestiges of the Late Heavy Bombardment could have persisted throughout geological time must find or devise an alternative explanation for each example.

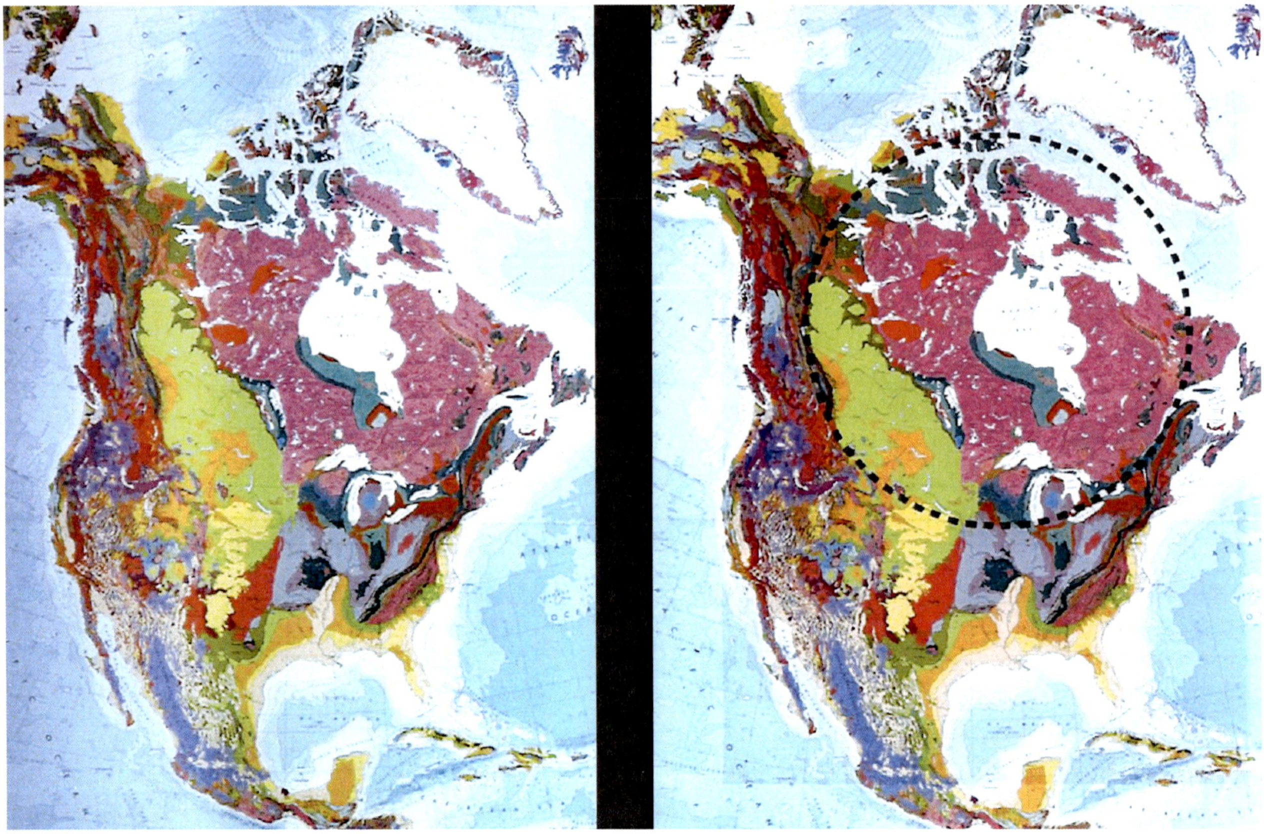

Figure 3.16 North America. Diameter of the circle approximately 3700 km. Base map: Geological World Atlas, 1:10,000,000, Sheets 2 and 3, ©C.G.M.W.-Unesco.

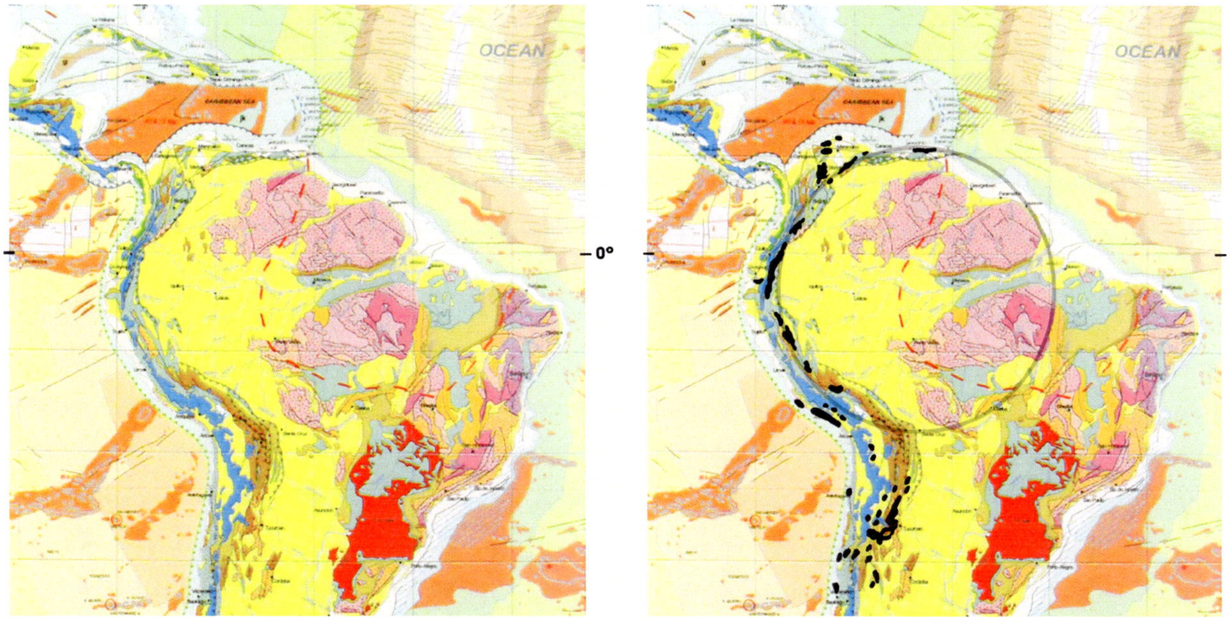

Figure 3.17 South America. The Guyanan and Brazilian shields are shown in violet. Slivers of comparably ancient rocks in the Andes are shown in black, their sizes greatly exaggerated for better visibility. Many of these slivers can be interpreted as fragments of the Guyanan and Brazilian shields, uplifted and incorporated in the Andes in the course of the ongoing Andean Orogeny (Hoorn et al., 2010). Other such fragments to the northwest of the circle were uplifted in what appears to have been different circumstances (see Hoorn et al., 2010). Modified from "Seismotectonic Map of the World – Five millennia of earthquakes around the world," 1:50,000,000 (First edition, 2001), Commission for the Geological Map of the World, Abdolazim Haghipour, general co-ordinator.

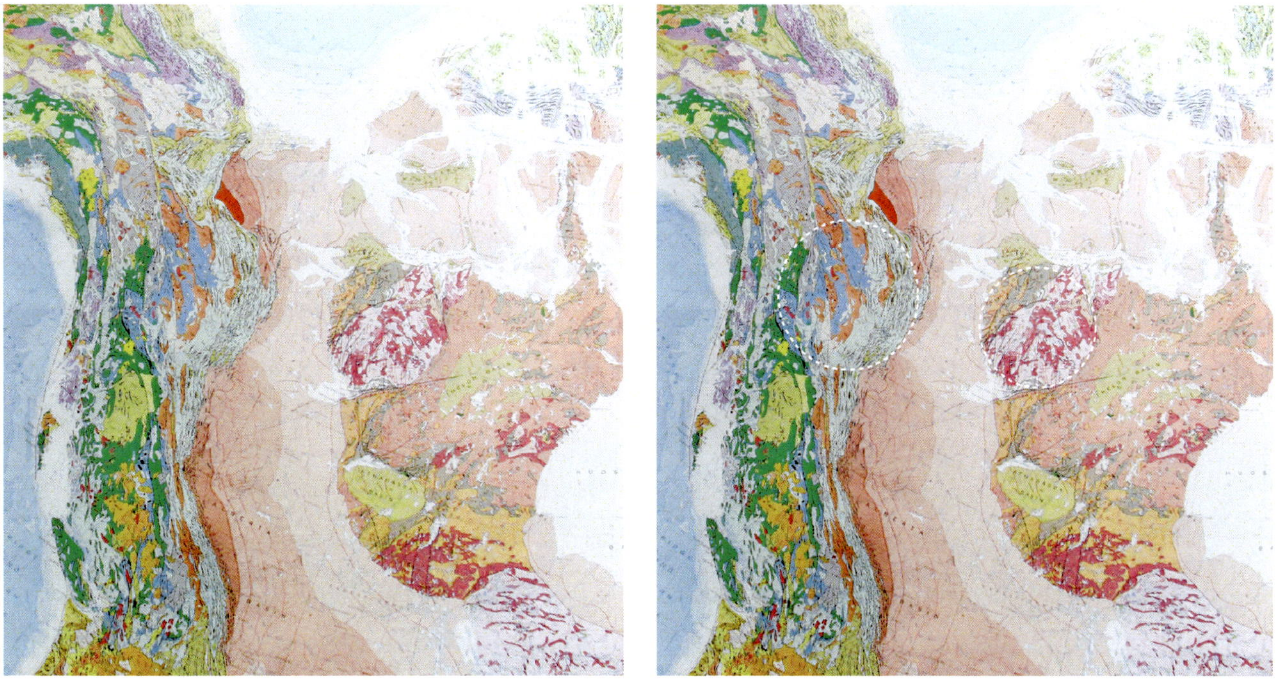

Figure 3.18 Two circular scars within the North American craton. Base map: Tectonic Map of North America, U.S. Geological Survey, 1:5,000,000, compiled by Philip B. King (1969).

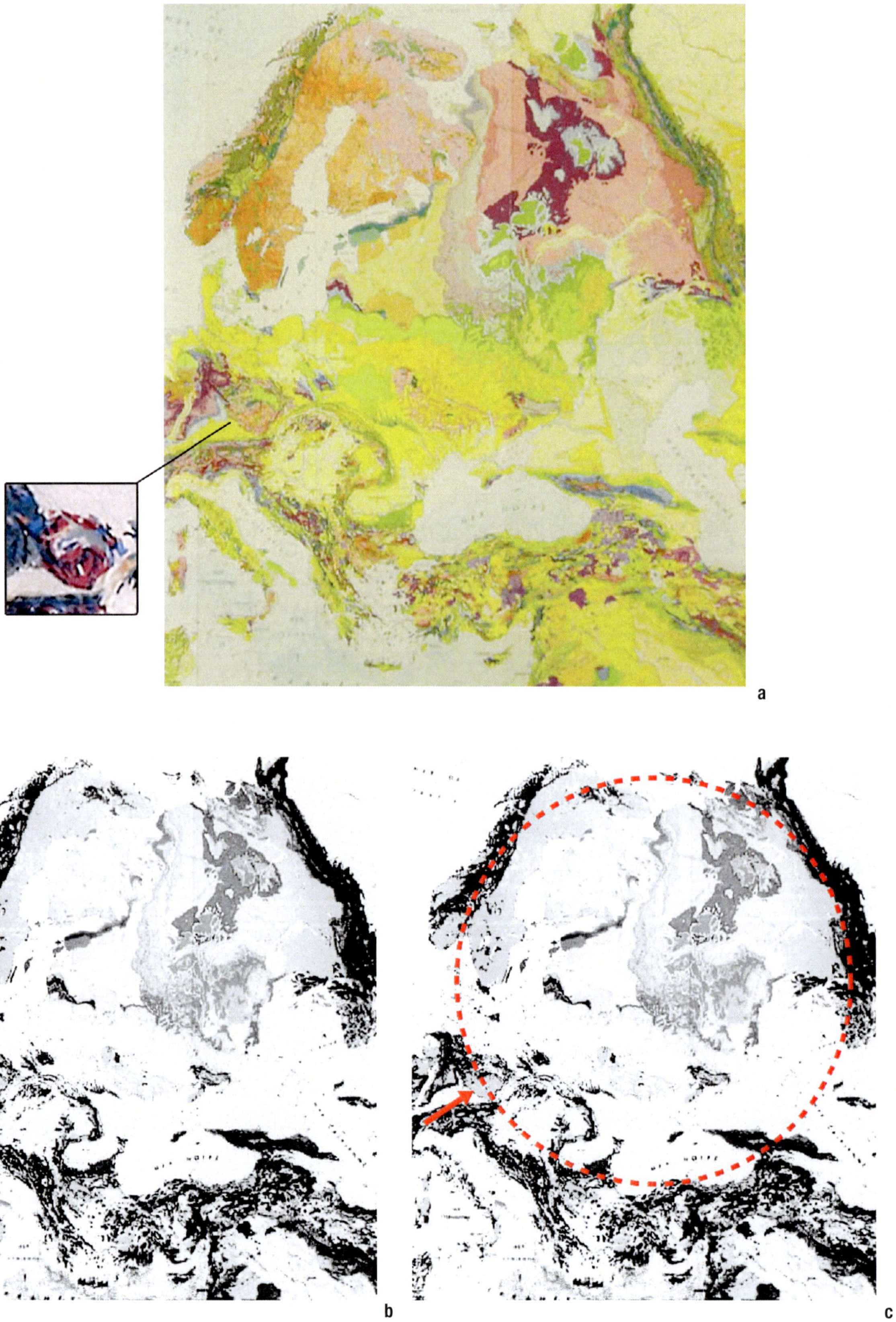

Figure 3.19a, b, c Versions of the geological map of Europe. Adapted from the Geological World Atlas, 1:10,000,000, Sheet 9. ©C.G.M.W. – Unesco. The inset shows the Bohemian Massif as depicted on a geological globe formerly at the Geological Museum in London.[57]

Much of North America has been scoured by glaciation, leaving the outline of the North American circle relatively easy to see. The circle is abutted on the west by the Canadian Rocky Mountains (formed 70 to 40 million years ago in the course of the Laramide Orogeny*) and on the southeast by the Appalachian Mountains (formed during the Acadian Orogeny, 375 to 325 million years ago).

With no reference to the Late Heavy Bombardment or to my own ideas, Keith Bloomfield, who worked for years as a geologist in Brazil, noted that "when information from drillholes is included, and the younger rocks of the Amazon Basin are ignored, the Guyanan and Brazilian Shields together form a nearly perfect circle."[56]

A number of smaller circular scars can be seen within the shields as in **Figure 3.18**, for example. I attribute these to late-arriving LHB impactors.

Each craton has had its own geological history. The North American craton has been scoured by glaciers; the South America craton is about half covered by the Amazon Basin and its sediments; and a large proportion of the ancient cratonic rock that defines Europe is overlain by diverse younger geological formations (**Fig. 3.19**). Each circle is difficult to see, or to recognize, but so are many other features of our planet.

The European circle is framed by three sets of mountains, each formed during a separate orogenic episode. The term *orogeny*, evoked in Chapter 1 as playing a critical role in the formation of transparent colored gemstones, refers to forces and events that led to large structural deformation of the Earth's crust and its uppermost mantle, the results of interactions between tectonic plates. Such interactions result in the formation of long tracts of highly deformed rock called orogens or orogenic belts. The craton circle that essentially defines Europe is bounded by the mountains of western Norway formed during the Caledonian Orogeny, some 490 to 390 million years ago, the Urals (formed about 318 to 252 million years ago during the Uralian Orogeny), and the Alps-Carpathians-Dinaric Alps-Greek mountains of the Alpine Orogeny formed about 55 million to 5 million years ago.

Orogens, which involve a great variety of geological processes, develop when a continental plate is crumpled to form mountain ranges. In Europe, plates were crumpled against the resistant European craton circle on three occasions (**Fig. 3.20**), much as the Andes were crumpled against the South American craton circle, and the Canadian Rockies and Appalachians against the North American craton circle.

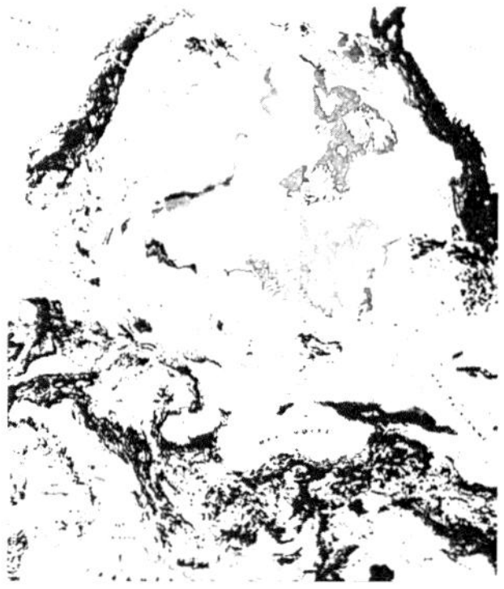

Figure 3.20 Europe. The circular shape emerges more visibly by using the artist's trick of backing away and viewing with half-closed eyes. Adapted from the Geological World Atlas, 1:10,000,000, Sheet 9. ©C.G.M.W. - Unesco

With the seeming exception of the Acasta Gneissic Complex of northern Canada[58] — a possible survivor of the LHB — the oldest known cratonic terranes are about 3600+ million years old, some 100 to 200 million years younger than the LHB. This interval may be a measure of the time required for "granitization."

Rocks on the platforms are younger still, formed by a variety of later geological events and processes. What the cratonic terranes and the platforms rocks have in common is an ancient broadly granitic basement that is relatively unfractured, far less so than the rocks of the orogenic belts.

Uplift at the edges of the cratons

At certain times and places along the fractured edges of the craton-forming scars, the Earth's crust would have been particularly weak, allowing light materials to rise by gravity, or heavier materials to be squeezed upward in the manner of toothpaste. The generally circular Bohemian Massif (inset **Fig. 3.19a,** and arrow **Fig. 3.19c**), which is still undergoing uplift, appears to be one such instance. Other otherwise unexplained instances of uplift include the Adirondacks[59] (with some 2000 meters of "excess" uplift) and the Black Hills[60] (with about a 1000 meters of unexplained uplift), perhaps squeezed upward during a tectonic collision (**Fig. 3.21**).†

* Orogenies, mentioned in Chapter 1 (*page 14*), are further discussed below.

† Dense rocks sink and lighter rocks are subjected to "gravitational uplift."

The Himalayas I: The "well-known" Himalayan arc

During much of the last 71 or more million years, the tectonic plate on which the peninsular portion of India sits has drifted northward (**Fig 1.46**). The initial collision with Mainland Asia involved the subduction of oceanic material trapped between the two continental masses. After the trapped oceanic material was subducted, inertia and other forces produced and prolonged the India-to-Asia continent-to-continent collision. In my view, this collisional history opened up part of a pre-existing LHB scar that then became markedly visible as the arc of the greater Himalayan ranges.

"How perfect is the Himalayan arc?" is the title of a study[61] in which measures were made of earthquake activity, changes in topography, stress in rocks, and kinks in the courses of streams draining southward from the mountains above. The results obtained in all four instances were statistically identical, giving an arc (**Fig. 3.22**), which, if extended, would form a circle 3392 kilometers in diameter with a margin of error of plus or minus 110 kilometers (3.25%).

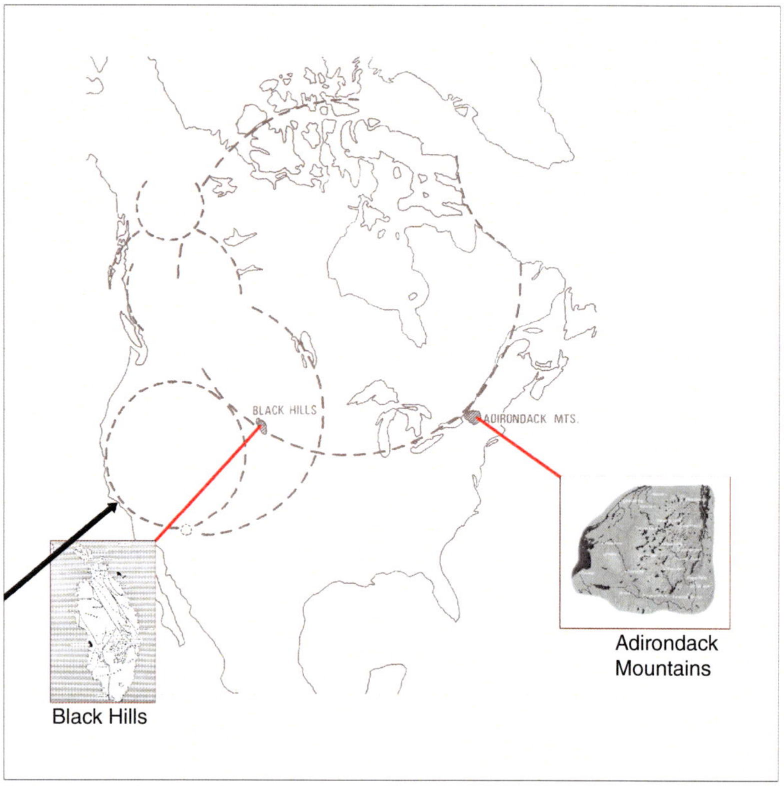

Figure 3.21 Canadian Shield, the Black Hills uplift, and the still-rising Adirondacks. The arrow indicates the direction of movement 74 to 64 million years ago of the Farallon plate (Bunge and Grand, 2000), now entirely subducted beneath western North America.

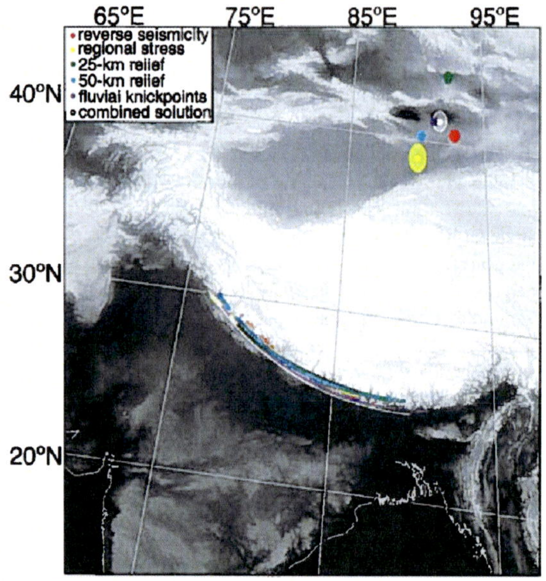

Figure 3. Best-fit small circles and poles with error ellipses. For central Himalaya, all small circles are statistically identical. We consider this result to demonstrate that (1) Himalayan boundary may be approximated as small circle with center of long 91.6°E, lat 42.4°N, and radius of 1696 km (combined solution), and (2) location of boundary is not sensitive to data type.

Figure 3.22 "How perfect is the Himalayan arc?" From R. Bendick and R. Bilham (2001) *Geology*, vol. 29, no. 9.

I have argued that circularity in itself is a characteristic that can be used in identifying terrestrial LHB scars. So from my point of view, it was unfortunate that the images of the quarter-circle of the Himalayan arc in the report "How perfect is the Himalayan arc?" do not indicate a complete circle.

Other images of the "well-known" Himalayan arc were seen in a variety sources, among them, part of an advertising poster spotted in the Paris Metro (**Fig 3.23**). Yet the full circular Himalayan scar that I presumed to exist was not apparent on any of the maps or globes consulted. Instead I saw a smooth curve terminated at both ends by a welter of confusingly complex topography, as in **Figure 3.23**. Northern portions of this hypothetical Himalayan circle appear to have been sealed by the collision itself, sealed too tightly to be recognized.

The extremities of the Himalayas are known as the "North West Himalayan Syntaxis" and the "East Himalayan Syntaxis," the word *syntaxis* being geological jargon for the convergence of mountain ranges.

Geological tradition includes the tried-and-true idea that no rock (meaning bedrock, not loose pebbles or boulders) got where it is by accident. The same notion is applicable to larger geological features such as mineral deposits and mountain ranges, and at this point we can ask why each end of the Himalayan chain should end with a syntaxis. It is assuredly not by accident.

The Himalayas II: The "unknown" Himalayan circle

The structure of the Himalayas is unique. This is obvious these days, though it was not always so. For although the greater Himalayan region is the only area currently undergoing continent-to-continent collision, this fact was not appreciated, nor could it have been appreciated, or even framed in a meaningful manner, until the late 1960s with the dawn of plate-tectonic theory. To the best of my knowledge, however, the theory of plate tectonics has not explained the existence of a syntaxis at each end of the Himalayan arc.

Matters became clearer in August 1984 in the course of a lecture in Moscow at the 27th International Geological Congress. A geological map of Asia was shown on which a much larger and more complete but "unknown" Himalayan circle some 5000 kilometers across could be easily seen and was captured in a now-faded snapshot (**Figs. 3.24** and **3.25**).

A scar 5000 kilometers across falls within the range of the expected. Back-of-the-envelope calculations based on the diameters of large impact scars elsewhere in the inner Solar System suggest that the diameter of the largest expected terrestrial LHB scar should be about that size.

The two Himalayan scars, shown as complete circles, are outlined in **Figure 3.26b** with the East Himalayan Syntaxis and the North West Himalayan Syntaxis at their intersections. Plugs of rocks with possibly circular outlines appear to have been uplifted in the vicinity of these particular parts of the Earth's crust, weakened by a double dose of deep fracturing.

Figure 3.23 Another image of the readily visible quarter circle of the "well-known" Himalayan arc. Additional circular patterns can also be discerned. ©1999 Face of the Earth.

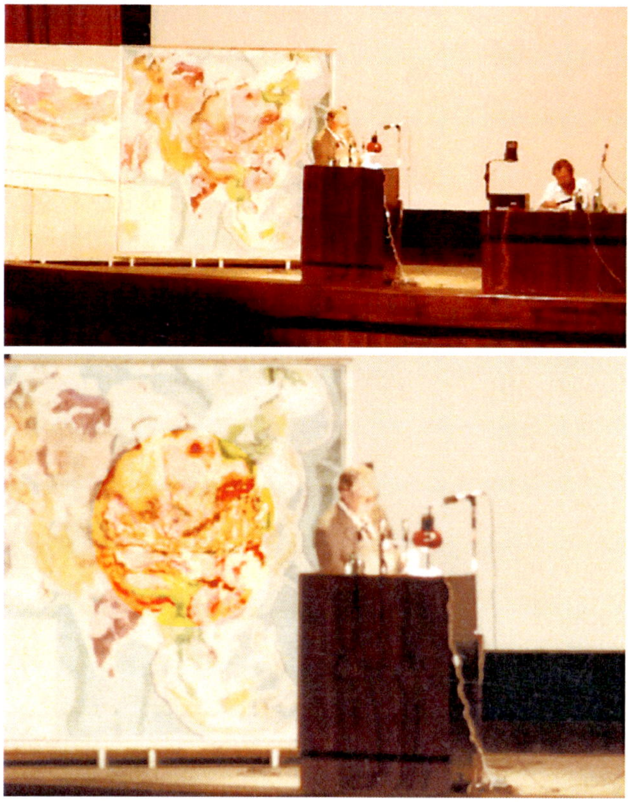

Figures 3.24 and 3.25 The "unknown" Himalayan circle; unenhanced and enhanced versions of a snapshot taken in a darkened lecture hall, J. Saul, Moscow, August 1984. Additional circular patterns can also be discerned. Novaya Zemlya, top center to the east of Scandinavia, is interpreted as the vestige of an outer rim. The map reference was not obtainable.

Rims

Many of the larger lunar impact sites show two or more rims, with pairs of rims separated by "moats."[63] Similar structures would not have survived on the Earth if they had been frozen ripple marks. (Ripple marks are "a folding phenomenon.") But if indeed caused by the inward flow of liquefied material followed by breakage ("a fracturing phenomenon") and the inward slumping[64] of the *entire thickness* of the brittle crust (*page 71*), outer rims of sufficiently large scars might survive.

Novaya Zemlya (**Fig. 3.27**) may be the vestige of a secondary rim of the larger Himalayan circle. Early investigators presumed that Novaya Zemlya was simply the northernmost part of the Urals, offset 600 kilometers to the northwest by the sharp bending of an originally continuous belt. This idea is now claimed to be incorrect,[65] for whereas the Ural Mountains were formed some 318 to 252 million years ago during the Uralian Orogeny, the folded structures of Novaya Zemlya are some tens of millions of years younger.[66] Although the age of mountain-building activity along similar belts — the Appalachians, for example — can

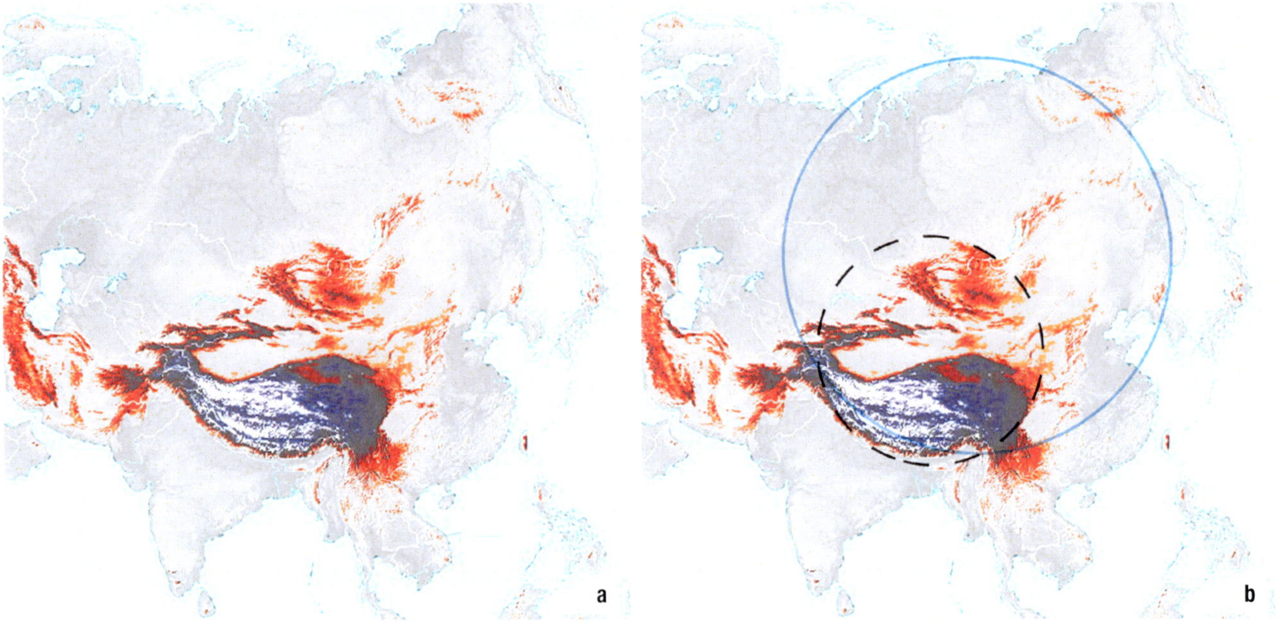

Figure 3.26a, b Himalayan scar circles; the intersections of the two circles closely approximate the position of the Himalayan front. Base map: *Physical Map of Asia*, http://geology.com/world/asia-physical-map.shtml

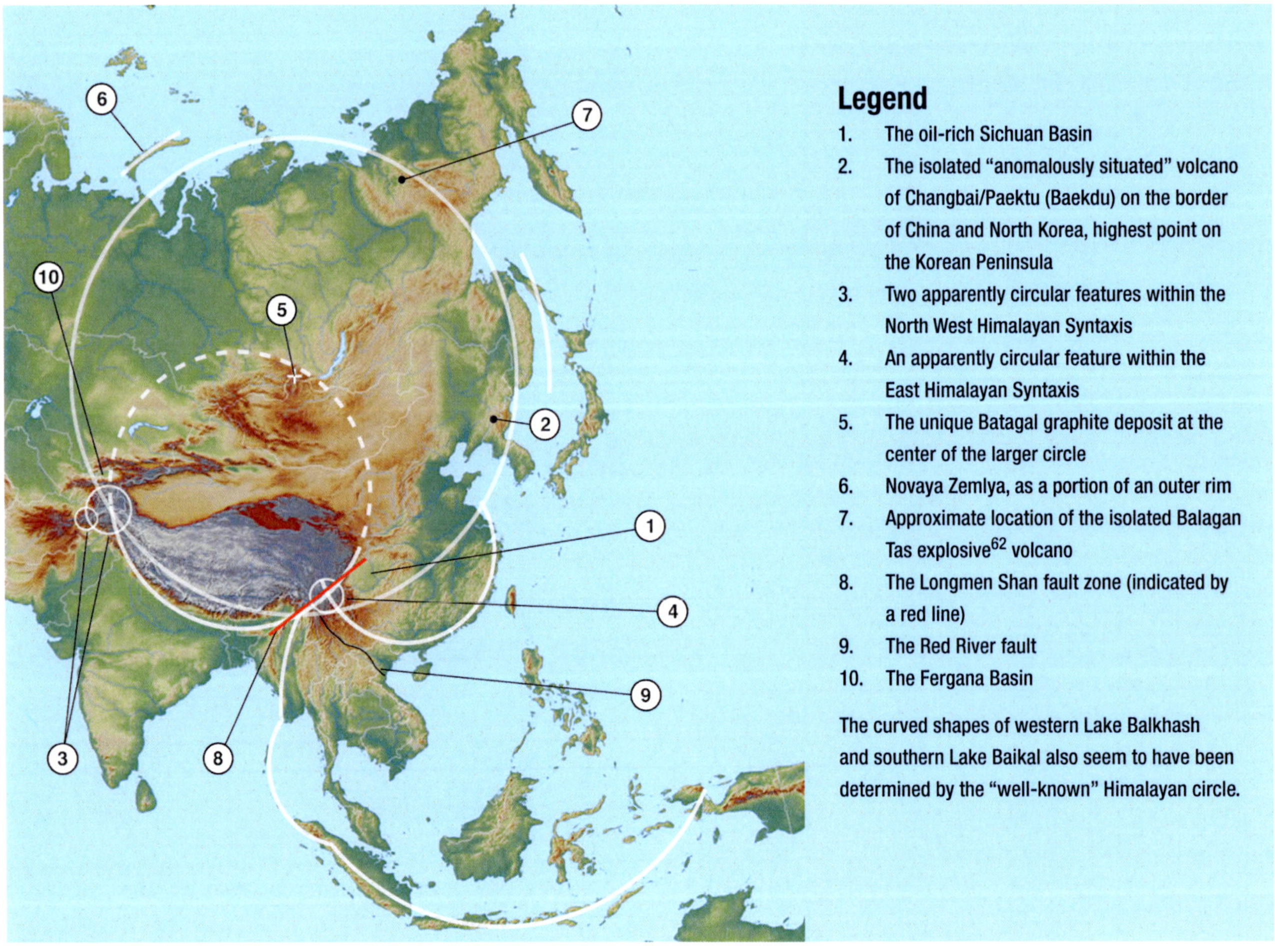

Figure 3.27 Selected geological features related to the two Himalayan circles. See text. Base map: *Physical Map of Asia,* http://geology.com/world/asia-physical-map.shtml

vary, the interpretation of Novaya Zemlya as an outer ring belonging to the great Asian scar seems in better accord with what is known of this remote area.

Elsewhere, the Adriatic and Italy may form parts of a moat and outer rim to the European circle (**Fig. 3.28**).

Europe

The circle defining the European craton abuts the greater circle of the Himalayas at the Urals where the continental plates of Europe and Asia collided during the extended interval of the Uralian Orogeny, some 318 to 252 million years ago.

These two cratonic units moved about, collided, built mountains, and lost material to subduction. And when they eventually halted about 252 million years ago, they possessed outlines that resembled those of LHB scars, rather than the irregular shapes of today's tectonic plates. This suggests that the outlines of some plates, or of some continental plates, are inherently circular. Material is accreted to their edges during some geological events, and is lost in the course of others, but the circle-shape essence remains.

The Uralian collision also trapped island arcs, mini-continents, and smaller circular units. In the Central Urals, the contact includes a roughly crescent-shaped area east of the Main Uralian Fault, a part-circle, shown outlined in white in **Figure 3.28**, in which are located the famous gemstone deposits of the Urals.

Additional circular scar patterns, not outlined in **Figure 3.28**, can also be discerned on this figure, especially in the region of Greece and Turkey.

Figure 3.28 Selected circular scars and rims, Europe. In the northeast, the European circle abuts an arc of the larger Himalayan scar. A smaller circle is shown outlined in white at the contact zone. Its northeastern portions, situated on the Asian side of the Main Uralian Fault, include the famous gemstone mining area of the Central Urals.

Rocks shown in pink in the circle north of the Black Sea are commonly referred to as the "Ukrainian Shield" but are essentially identical to the shield rocks of Finland and Scandinavia, also shown in pink. Volodarsk (Korosten), situated within this area and mentioned in the previous chapters, is notable for its production of large gem-quality topaz and beryl. Geological World Atlas, 1:10,000,000, sheet 9. ©C.G.M.W.-Unesco

At the center of Asia

On the Moon and on other planetary bodies, the floors of certain large complex LHB impact sites include central peaks. These are believed to have a twofold origin, caused by rebound, and also by symmetric forces exerted inward immediately after impact as the crater walls collapsed, pushing up the central part of the floor from all directions. Central peaks surely existed on Earth following the LHB, but their survival into present times is uncertain.

The most promising candidate site for an LHB peak is at the center of the larger Himalayan circle at Batagal (East Sayan, Siberia; marked + and ⑤ on **Fig. 3.27**).

Batagal formerly produced graphite of uniquely fine quality, which, in common with ivory, could be carved in exquisite detail **Figure 3.29**.

The source of carbon at Batagal is reputed to be the very deep crust or the Earth's mantle, supposedly from an alkaline plume dated 520 to 460 Ma that "carbonized" metamorphic rocks with graphite, carbonate, and bitumens. Platinum and palladium metals are also present at Batagal, as is pure silicon, an exceedingly rare occurrence in nature. Ophiolites, serpentinization, inclusions with methane and nitrogen, and

Figure 3.29 Block of graphite from Batagal, 52°21′ N, 100°45′ E, East Sayan, Buryatia, Siberia, Russia. The fly on the polished surface of the upright block is carved in graphite and is approximately natural size. Photography by F.L.

an alternative age of carbonization around 690 Ma are also reported, as are supposed bacterial remains.

Despite more than 150 years of investigations, the genesis of the Batagal deposit remains disputed, a highly confusing and unique mineral deposit, located in a very particular geographical spot.[67]

Surviving LHB scars not associated with the formation of cratons

In the course of the Late Heavy Bombardment, a great number of lesser impacts would have also penetrated the entire thickness of the Earth's crust, producing substantial quantities of melt, though not enough to lead to the formation of cratons. On solidifying, this melt would have provided relatively impermeable circular plugs surrounded by narrow 3-D craterform boundaries composed of rocks that had been severely fractured and broken up by the impact, but not melted.

Such scars extended — and, I claim, must still extend — down to the boundary where brittle rocks cease to exist and rocks are ductile.* Below this boundary, materials may remain essentially solid but are able to act in ways that resemble an extremely viscous fluid, even to flow.† Some man-made materials flow slowly when gently tipped but fracture or shatter when hit, and pitch exhibits similar properties. No drops fell from a pitch-filled glass funnel in an exhibit in the science building of my school during the four years I was there, but two drops had fallen by the time I returned for my tenth class reunion.

Matters did not rest, whether figuratively or literally, once an impact penetrated to the ductile domain. The Earth is geologically active. Rising fluids, earthquakes, and the twice-daily "Earth tide" provide constant jiggling. (The magnitude of the Earth tide, which can be as great as 55 centimeters (20+ inches) at the Equator these days and was greater in the past, is not affected by the rigidity of rocks.) In consequence, an LHB or other fracture that attained the depth of the brittle-ductile boundary would be intermittently propagated and repropagated upward into and through any new surface rocks, a matter of *inheritance from below*. Such fracture patterns might never be able to heal[68] and might have persisted throughout all geological time, though not necessarily readily discernible at the Earth's surface at any particular geological moment.‡

In common with a smile or a frown, fracture-formed circular scars visible on the surface of the Earth are indications of something deeper. And also in common with a smile or a frown, they generally require critical interpretation. In places, the fracturing and permeability along the edges of such scars would have provided conduits for many sorts of fluids. They would provide elements of the Earth's "deeply penetrating fundamental 'plumbing' system,"[69] guiding the penetration and flow of surface waters, groundwater, primordial fluids, gas, petroleum, magma, lava, mineralizing fluids, and barren fluids, as well as heat, whose flow obeys the same physical laws as conventional fluids.

The perimeters of scar circles that are too small to be associated with cratons are in places marked by mountains, valleys, rivers, certain types of mineral deposits, boundaries between rock types, changes in vegetation, gas vents, hot springs, volcanoes, tabular bodies of igneous rocks ("dikes"), tracks and trails, and roads and railroads whose routes followed paths of least resistance. Alternatively, the outlines of such scars may be temporarily masked

---

* The depth of the brittle-ductile boundary varies from place to place, and its average, minimum, and maximum depth, either at the time of the LHB or now, are unknown. Estimates of various sorts run from less than 2.5 km to more than 40 km. With subduction, slabs of cold, brittle rock descend far below the normal brittle-ductile boundary. Deep earthquakes occur within descending slabs as increasing temperature and pressure cause them to expel water and other volatiles.

† Such flow is termed "creep" and involves slow movement that produces permanent deformation. Creep may be the consequence of high stress that remains below the yield strength of the rock, and it may involve the sliding of crystals past one another.

‡ A counterargument here is that ductile rocks generally flow, that such flow involves shear, and that the shear obliterates original features. This argument is surely valid beginning at depths somewhat below the brittle-ductile boundary, but far less so, or not at all, at the boundary itself.

or covered by lava, sediments, or lakes, or they may be shut tightly and essentially invisible due to compressional stress, as may be the case for the northern portions of the "well-known" Himalayan circle.

To summarize, the rim zones of impact scars dating to the Late Heavy Bombardment are characterized by deeply penetrating fractures that provide paths of lessened resistance and are perennially or sporadically available as conduits for Earth fluids. *These patterns are far older than the rocks in which they are observed.*

Terrestrial LHB scars and mineral deposits

Fractures on the rims of LHB scars have acted as conduits for fluids now and then throughout our planet's history. Particularly clear examples come from a mining district in Arizona that has produced much of the copper and gold mined in the United States. There, the main deposits (known as porphyry deposits) are characterized by fractures, veins, veinlets, and severely broken-up and re-cemented rocks (breccias);* (**Fig. 3.30**). The origin of the fracturing has been much debated, but one study from the 1970s concluded that "some of the breccias and adjacent rocks in the vicinity of porphyry deposits ... have features that seem indicative of shock metamorphism."[70]

Figure 3.30 Colorful breccia from the Emma Luisa porphyry deposit, Huanaco (Guanaco), Chile. The bright blue mineral is ceruleite, a rare water-bearing copper aluminum arsenate. Photo: J. Saul

Porphyry deposits are found in regional clusters in which ancient fracture zones have been opened and reopened by high-pressure mineral-bearing fluids in a sort of natural fracking process. The metal-bearing minerals are deposited in pulses when and where the pressure and temperature of the fluid become sufficiently low, and are further concentrated by interactions with circulating groundwater.

The topography of the mining district in Arizona is shown in **Figure 3.31**. Circular patterns visible in the same area are highlighted in **Figure 3.32**, and **Figure 3.33** shows the two images combined. Small numbers in **Figures 3.32** and **3.33** indicate porphyry-type deposits and the few other mineral deposits in the area whose existence is also due to rising or descending fluids. Total ore reserves of this area as of 1973, the year the information was published,[71] were worth many billions of dollars.†

No other such mineral deposits are known from the region. The blank areas in **Figures 3.32** *and* **3.33** *are indeed blank*, lacking readily discernible circular patterns, and also lacking mineral deposits (known as of 1973 when the data was compiled).[72]

The correlation of mineralization with circular patterns in this part of the world was first recognized in 1974, and first published in 1978.[73] It involves the systematic association of valuable mineralization

- with readily discernible circular patterns,[74]
- with breccias,
- with the flow of fluids,[75] and
- with a zone in which the Earth's brittle-to-ductile transition is suspected to be particularly shallow (the Arizona Transition Zone[76]).

* These characteristics of the rock allowed fluids to circulate and thereby concentrate metals that were initially too dispersed to be economically mined.

† Metals and other materials extracted from this area include gold, silver, copper, lead, zinc, molybdenum, tungsten, manganese, iron, vanadium, turquoise, and serpentinized limestone.

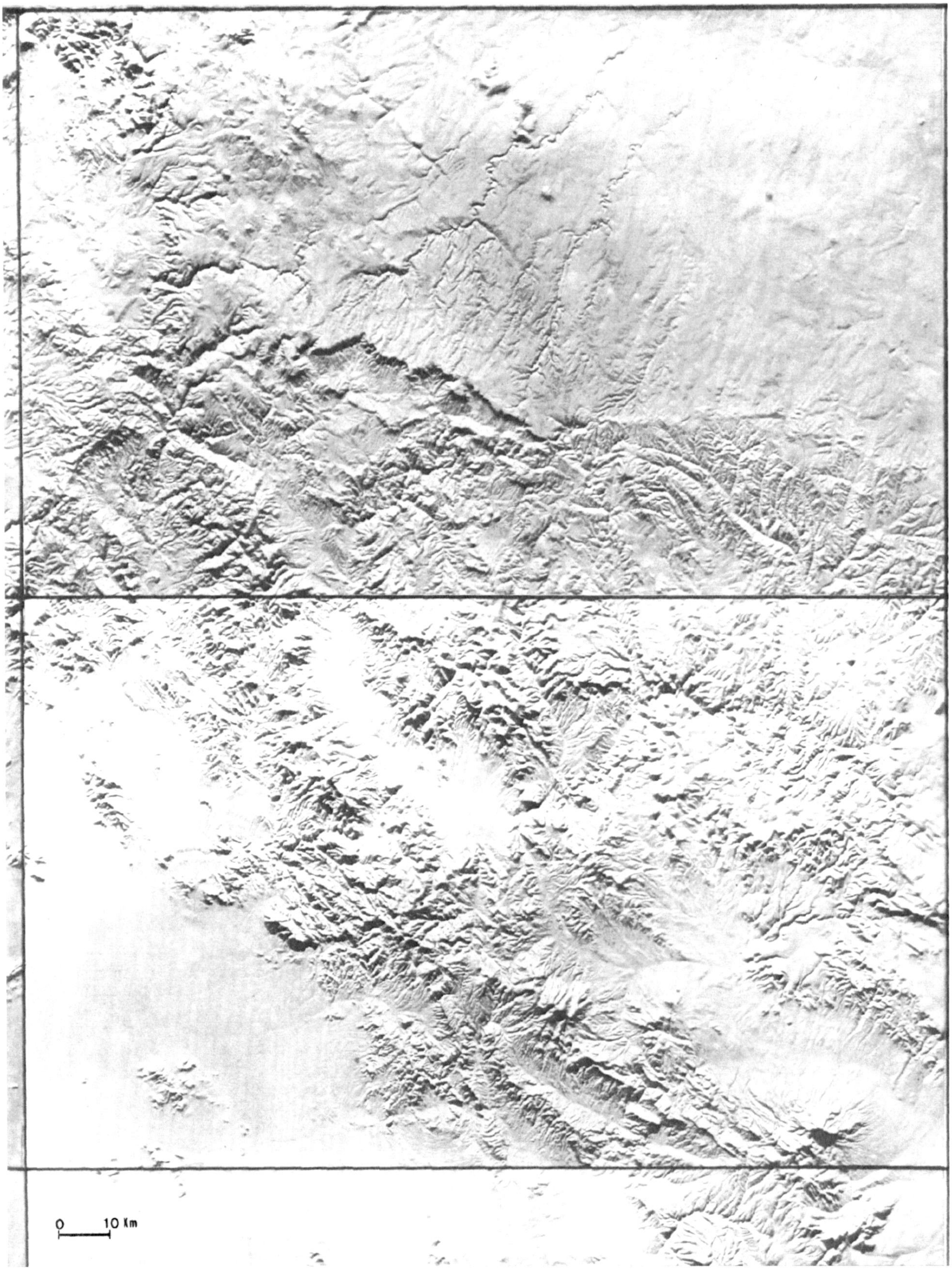

Figure 3.31 Portion of southwestern Arizona showing topography (adapted from Saul, 1978). The largest circular scar here corresponds to the very small circle in Figure 3.21. Part of the ill-defined Arizona Transition Zone traverses the image from northwest to southeast.

Figure 3.32 Portion of southwestern Arizona showing outlines of circular scars and the locations of mines (as of 1973) in which deposition or concentration of the ore involved the circulation of fluids (adapted from Saul, 1978, in which each number is keyed to a named mine[77]). The largest circle here corresponds to the very small circle in the southwest of Figure 3.21.

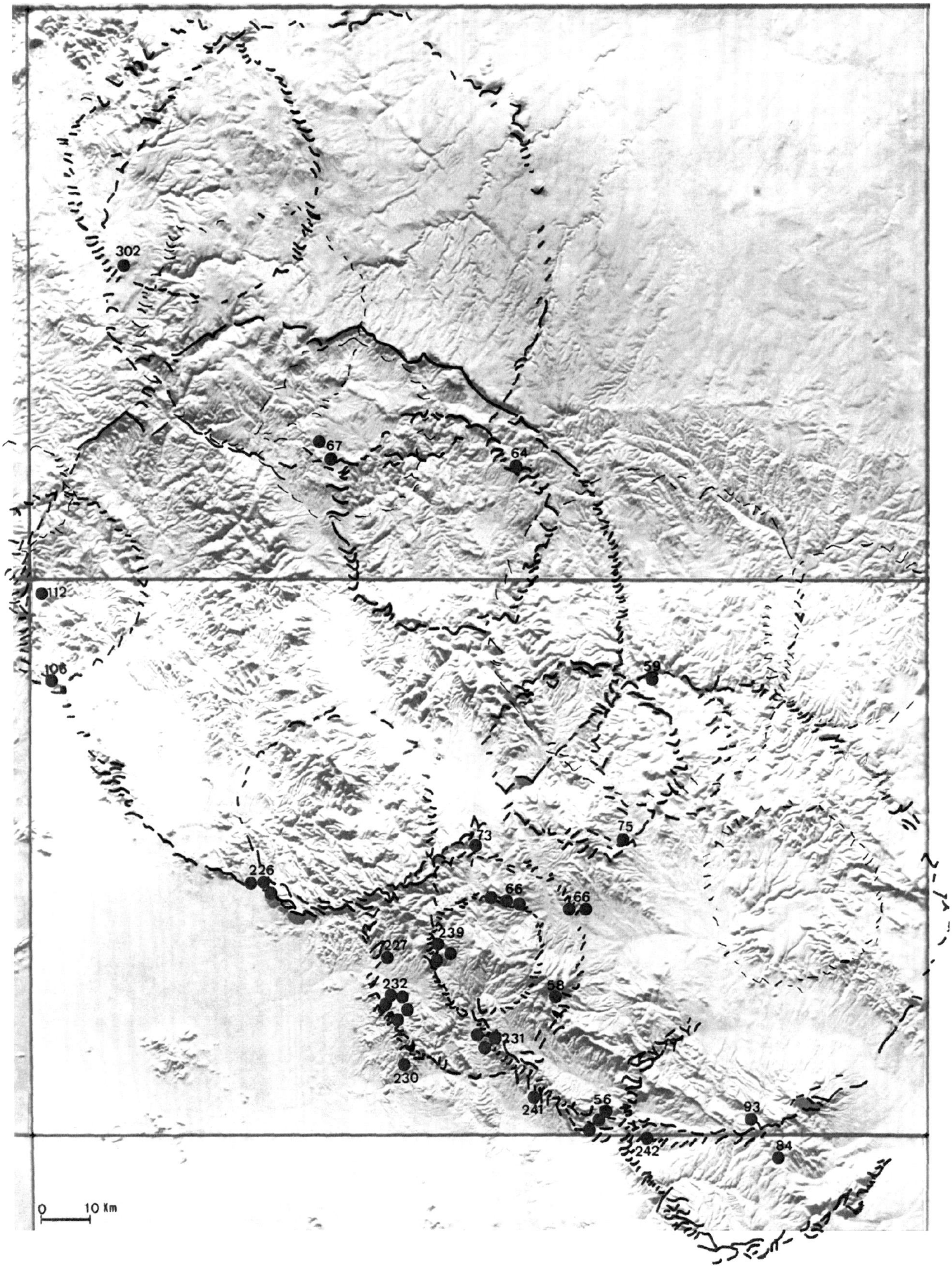

Figure 3.33 Superposition of the outlines of circular scars on the topography of a portion of southwestern Arizona, made by combining the two preceding figures.

Replies to frequently asked questions

- Many LHB scars are difficult see. If it had been otherwise, they would have been investigated long ago.
- The scars are very numerous. This is as expected by analogy with the Moon.
- *All* of the circular patterns are far older than *any* of the observable geological features that today define them. Most of the gross features of our planet are inherited, consequences of initial conditions established by the Late Heavy Bombardment.
- The scars do not resemble great volcanic features such as Mauna Loa, the Yellowstone Caldera (55 km x 72 km), or Indonesia's Lake Toba (100 km x 30 km). Nor should they.
- Dents in ping-pong balls also produce arcs with circular curvature. No one assigns the circular features on the Moon to the dent-in-a-ping-pong-ball effect.
- One of the principles of modern geology holds that the features of our planet were formed in times past by much the same forces as we see at present: winds, currents, sedimentation, volcanic activity, and so on. "The present is the key to the past." "Uniformitarianism," as it is called, has been a prevailing paradigm for English-speaking geologists ever since the publication of Charles Lyell's *Principles of Geology* (3 volumes, 1830–1833). The opposing view, which evokes great upheavals and floods such as that described in Genesis, has been derogatorily called catastrophism and was in great disfavor for the century and a half following publication of Lyell's *Principles.* This situation has been changing since the 1980s. But as a consequence of their training, many established geologists will greet the idea that a "catastrophic" bombardment has established the initial conditions of geology as inherently suspect or tainted.

Decades ago Pierre Nicolini, a specialist on the formation of ores, observed that secondary branching faults are more commonly mineralized than major regional faults.[78] His counterintuitive contention received virtually no attention at the time but was recently explained by Dion Weatherley and Richard Henley, who noted that whereas the sides of an active fault move past one another during earthquakes, lesser fractures that intersect the main fault react differently, opening up at the beginning of the quake. Any mineral-bearing liquids within such fractures then undergo a great drop in pressure, perhaps over a thousandfold. They may then vaporize and instantaneously "lose their load," depositing small quantities of silica (quartz), gold, or other materials by "flash deposition." In time the secondary fractures may refill, and the process repeat with further deposition. An intriguing aspect of Weatherley and Henley's work is their calculation that flash deposition occurs even during small earthquakes, provided that the secondary cracks have had time to be refilled.[79] It is exactly these circumstances — small earthquakes and rapid refilling — that might be expected along the fractured perimeters of certain circular scars.

Figures 3.34–3.36 show what I call the Stockholm Circle, a geographical feature frequently noted but perhaps never attributed to the Late Heavy Bombardment before 1981.[80] The locations of metal deposits of various sorts in the northwestern portion of the circle are indicated in **Figure 3.34**, as well as the orientations of ore-bearing strata or veins, many of which are generally tangential to the circle. Contacts between one rock unit and another on islands of the Stockholm Archipelago are also tangential to the circle. The original maps are in color, but the circle is more easily appreciated in smaller copies in tones of gray.

The inheritance of deep fractures

Old-timers who have worked at a particular mine for decades commonly insist that the same fault zones had provided pathways for ore-bearing fluids time and time again. Comments by various specialists follow, all saying essentially the same thing.

> ...the geochemical permanence in certain provinces and certain aspects of the large mineralized linear features... can only be explained by successive inheritances right through all the [geological happenings] which have occurred since very distant geological times.
>
> —Pierre Routhier, "Sur trois principes généraux de la metallogénie et de la recherche minérale" (1968) p. 217.

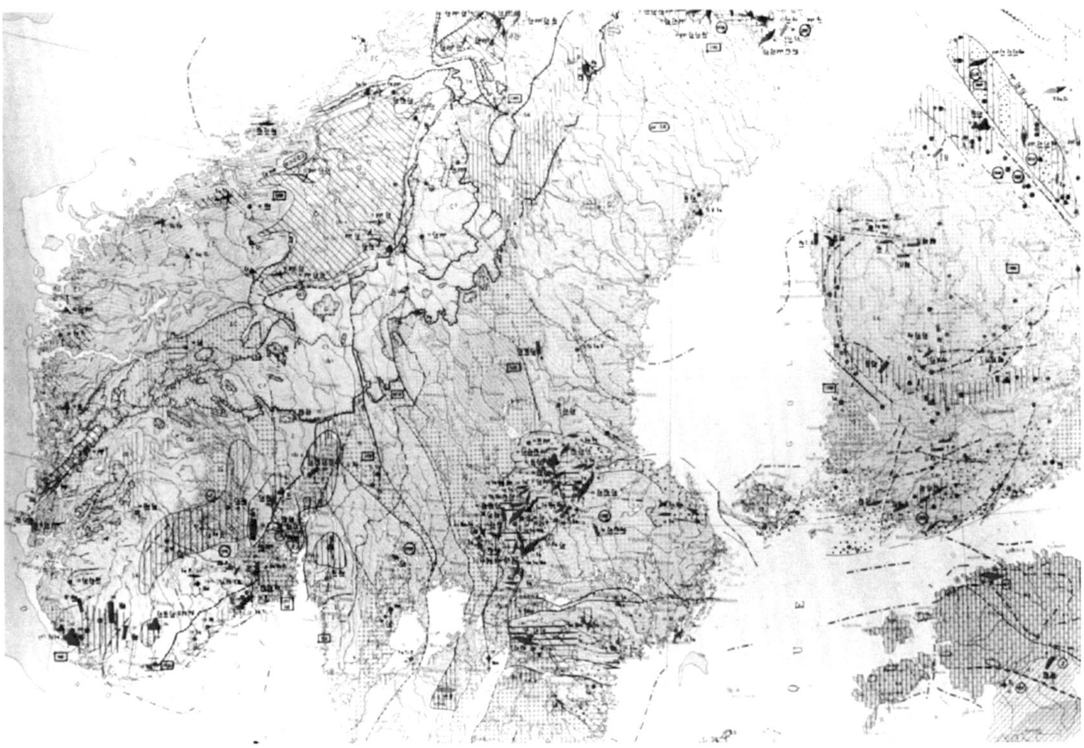

Figure 3.34

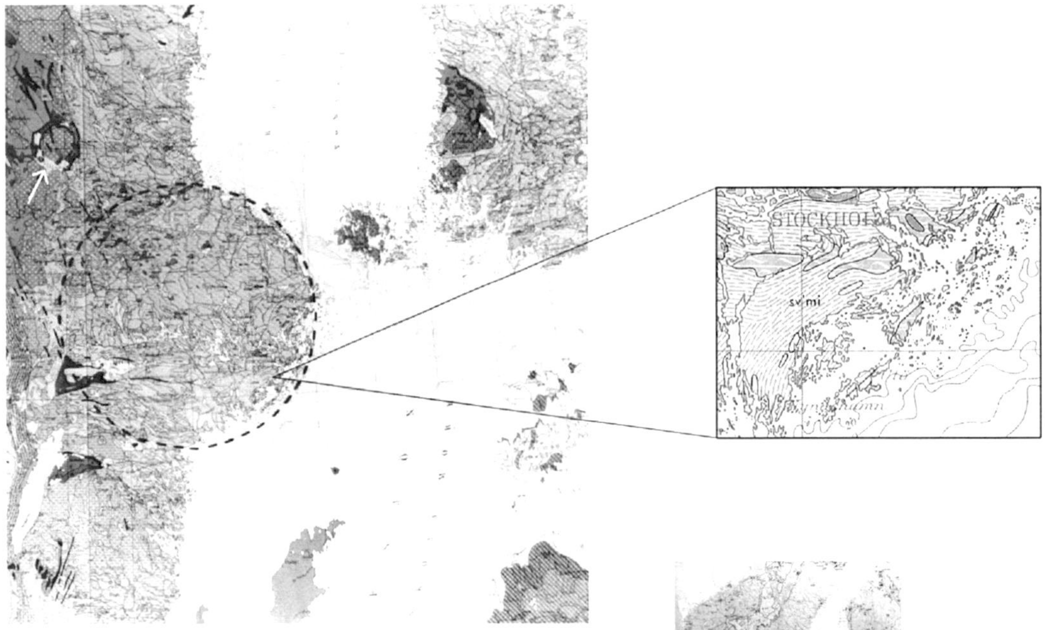

Figure 3.35

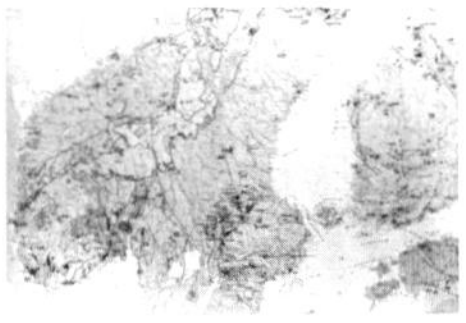

Figure 3.36

Figures 3.34, 3.35, and 3.36 Maps of the Stockholm Circle. Figure 3.34 shows occurrences of metal deposits, and Figure 3.35 is a geological map with details showing geological faults and contacts along the perimeter of the circle. The small Fig. 3.36 shows an unenhanced but greatly reduced version of Figure 3.34, devoid of distracting detail, on which the circle is particularly easy to see. The smaller circle indicated by an arrow in upper left of Figure 3.35 is the approximately 377 million-year-old Siljan Ring, 52 km in diameter, and one of the 185 accepted post-LHB meteorite impact sites. Sources: Carte métallogénique de l'Europe, 1:2,500,000, Sheet 2 – Europe N (1968–1970) ©Unesco and Bureau de Recherches Géologiques et Minières; and Carte géologique de l'Europe, 1:500,000, Sheet D3, 2nd edition (1966) ©Bundesanstalt für Bodenforschung / Unesco.

...one of the features of concentrically annular structures and breccia zones is their longevity and inherited character of development along with ... high ... permeability. A second feature is conservation of their morphology despite the material transformations of their constituent rocks.[81]

—M.Z. Glukhovskiy, "Ring Structures and Linear Faults in the Aldan Shield and Stanovoy Region" (1978) p. 331.

...confirms once more the extremely great tectonic mobility and permeability of the outer zones of circular systems, which persisted during subsequent geological periods.

—V.M. Moralyov and M.Z. Glukhovskii, "Circular Features of Precambrian Shields as Recognized on Space Photographs," *Soviet Journal of Remote Sensing*, no. 3 (1981) p. 359.[82]

...preexisting fractures can control the behavior of rocks, even at depths where no voids remain.

—Richard A. Kerr, *Science* (27 Nov. 1987) p. 1227.

... permanent zones of weakness which acted more or less continuously from Precambrian to present.

—A. Piqué, M. Dahmani, D. Jeanette, and L. Bahi, "Permanence of structural lines in Morocco from Precambrian to Present" (1987) p. 247.

Another curiosity is the obvious repetition of mineral deposition in the veins.

—Richard A. Kerr, *Science* (8 January 1988) p. 145.

"Repeating Deep Earthquakes: Evidence for Fault Reactivation at Great Depth" - title of an article in which some deep earthquakes are reported to be "colocated" in 3-D.

—Douglas A. Wiens and Nathaniel O. Snider, *Science* (24 August 2001) p. 1463.

Big trouble for acceptance of the circular scars

Reporting in the *New Scientist* of 11 August 1977, Keith Hindley,[83] who had attended a presentation I made at a meeting of the Meteoritical Society in Cambridge the previous month, wrote that "The more obvious features are certainly real and seem to represent fracture systems because they correlate with mineral deposits. ... The evidence is very suggestive but the rings are faint and Saul is going to have problems convincing the skeptics. As he put it, the rings are the smile of the Cheshire cat rather than the Cheshire cat itself."

A standard sign at railway crossings in France warns drivers and pedestrians that one train can be concealed behind another. Similarly, it turns out that a second problem for the acceptance of my circles work lies behind the practical difficulties in observing the ancient scars. Hindley was entirely right about the skeptics in 1977, but the problem no longer concerns the faintness of the circles or the fact that parts of their perimeters are commonly masked. Indeed, when I gave a similar presentation to some 70 students years later, one complained that I kept showing examples long after she had been convinced.

The problem these days is the apparent clash with plate-tectonic theory. In my presentations, I did not only associate terrestrial LHB scars with cratons and mineralized circles on the continents. I also discussed subduction zones, and the subduction of oceanic crust. Trouble arises in peoples' minds from the fact that oceanic crust is constantly consumed by subduction and renewed at mid-oceanic ridges, and that there is no oceanic crust older than 200 million years. "Therefore," or so it is claimed, the Earth's oceanic crust cannot possess 3800+ million year old scars.

But the argument should be turned around. In their 200 million years (at most) of existence, the basalts and underlying mantle would have moved far from the spreading ridge while becoming progressively colder and more dense. At some point, they would ride over an arc that had become significantly weakened from below and broadened by the regeneration of fractures dating to the LHB. The regeneration need not have reached the actual surface, only the vicinity of the surface. The oceanic plate might then break through and subduction might begin. The regeneration need not, and would not, have reached the Earth's surface.

In times past, arcs may have provided the only places suitable for the initiation of subduction. These days subduction also occurs along deep linear fracture zones. Some of these linear zones appear to be tangents to circles; others may be paths of least resistance, deep breaks connecting one circular or linear scar to another.

Subduction

Without subduction zones, there would be no plate tectonics.[84] Yet as matters stand, the prime shortcoming of plate-tectonic theory concerns the unknown manner by which the first deep subduction event was initiated. How did deep subduction first start? When did the first plate subduct? Why not earlier and why not later?

Two factors have been evoked: cooling of overlying materials and their concurrent increase in density, and weakening of underlying materials. Even combined, however, they fail to fully address a critical matter. For in contrast to erosion, sedimentation, crystallization, or the generation of melts, deep subduction has not been a geological process throughout all geological time. The much-debated oldest vestiges of deep subduction are apparently only about 500 to 1000 million years old. To understand the origin of *deep* subduction as a geological process (and the introduction into the Earth's mantle of great quantities of seawater that lower the melting point of mantle materials and engender partial melting), I propose to revisit both factors, the cooling of the overlying basalt and the weakening of the underlying material.

Cooling of overlying materials

In a particularly useful review article published in 2007,[85] Robert J. Stern argued that the early Earth had been hotter than it is now and that slabs of oceanic-type rocks (basalts) would have been less dense than comparable slabs in more recent times. Their elevated temperature, Stern argued, would have made them too buoyant to sink.* He concluded that deep subduction could not have occurred until the Earth had cooled down and hence, that "plate tectonics on the early Earth occurred sporadically if at all."[86]

After acknowledging undeniable evidence for subduction-like phenomena in early times, Stern made a distinction between "modern-style plate tectonics" with deep subduction and whatever came before. The "whatever came before" apparently included occasional, or perhaps frequent, shallow, sluggish subduction that reactivated nearby cracks and faults without, however, displacing tectonic plates through substantial distances.[87] Stern calls this "stagnant lid tectonics."

These days, Stern notes, slabs of rock become candidates for subduction when the difference in density between the slab itself and the underlying material exceeds 1½ percent. A difference of this magnitude produces enough "negative buoyancy," he observes, to overcome the strength of the underlying material. This "strength," however, would not be a typical or average strength of the underlying material. Rather, it would be the strength of a particularly weak portion of the Earth's crust.

Weakening of underlying materials

During and following the Late Heavy Bombardment, major landforms on the Earth's surface would have included great unstable soil-free crater walls in the process of collapsing under the force of gravity and undergoing the effects of chemical weathering and physical erosion. Shattered rocks are highly permeable and especially subject to erosion, and what had once been a craterwall mountain could eventually become a valley.

* The underlying material would have also been hotter, but the difference in density would have been less, and it is the difference in density that determines whether subduction can occur.

After one or two or three billion years and numerous cycles of erosion, sedimentation, uplift, and more erosion, the fractured rim zones of some LHB scars would have been exposed to substantial depths. **Figure 3.37a** shows a schematic profile (cross section) of a complex LHB crater, the thick line indicating the boundary of the transient crater formed during the impact itself, i.e., the boundary between materials that had undergone liquefaction and the less-affected rocks that had only undergone fracturing. The thin line represents the much shallower resultant crater that was formed following rebound of the liquefied rock. The dotted square has been added to indicate the angle — 66° in this particular example — at which the transient crater had intersected the Earth's surface.

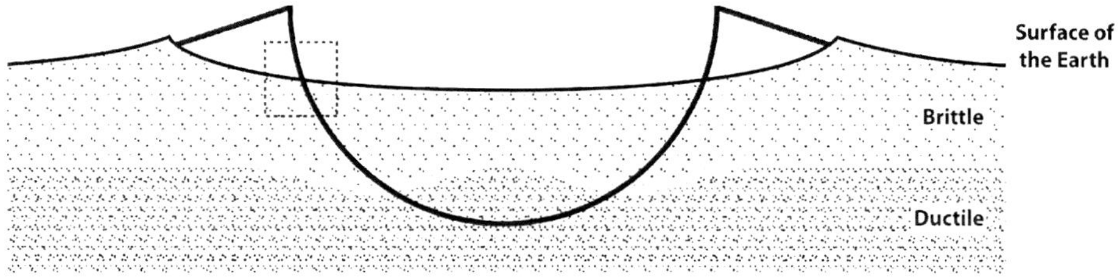

Figure 3.37 Schematic cross sections of a large, complex impact crater showing its evolution in the course of erosion. Overriding tectonic plate is not shown. See text.

a

Extensive erosion, **Figure 3.37b**, and still more erosion, **Figure 3.37c**, exposes lower levels of the transient crater. As can be seen, the angle at which the transient crater intersects the Earth's surface is progressively lower from **Figure 3.37a** to **Figure 3.37c**.

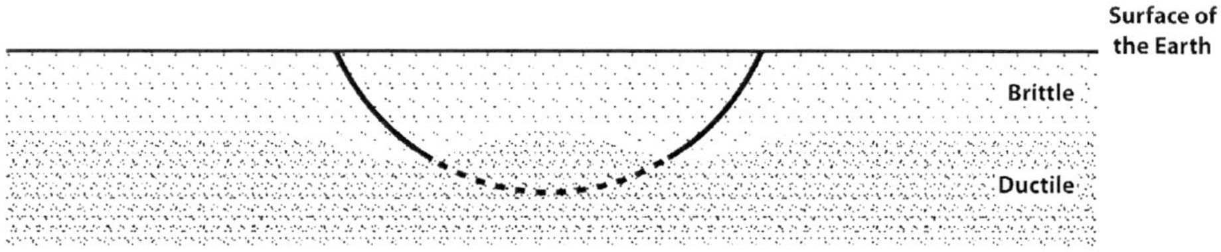

b

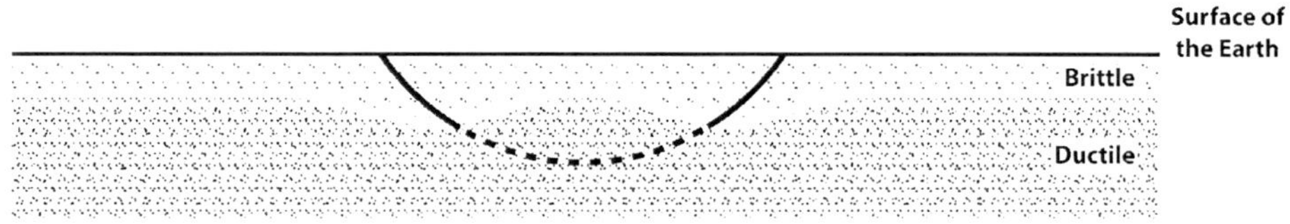

c

Figure 3.37d shows the remains of the same crater at a still later time: the site has now been covered by a new generation of overlying rock.

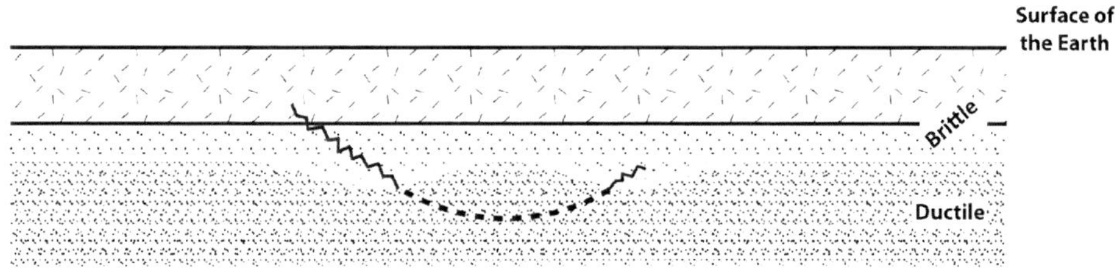

d

Fractures propagate in straight lines, and the regenerated fracture will eventually cut the surface of the Earth at a much lower angle, 33° in this particular example (as in the dotted square in **Fig. 3.37e**), than did the top of the transient crater immediately following the impact (dotted square in **Fig. 3.37a**).

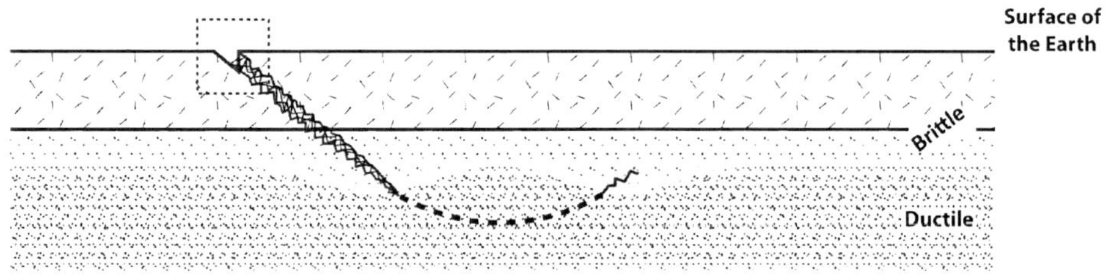

e

As mentioned earlier, ductile rocks flow, such flows involve shear, and shearing would be expected to obliterate original features. This might well be the case at depths below the brittle-ductile boundary,[88] but would be far less so, or not at all, in the immediate brittle-ductile transition zone.

Figures 3.37a to **3.37e** illustrate a sequence of events during which fractures from the transient crater were repropagated upward. These schematic figures are simplifications; there would have been numerous cycles, during which the original fractures were reactivated at angles that were progressively less and less steep. The result would have been a broadened fanlike fracture zone, widest at the top and with a width of, say, a river valley.

A gently descending zone of fractured rock tens or hundreds of kilometers across could allow a slab of denser material to slide into and underneath it, to subduct.* Various forces may be invoked, but gravity would be the prime consideration. I would hope that this explanation would have satisfied the great Soviet geologist, Vladimir V. Beloussov (1907–1990), one of the last authorities to reject plate tectonics, disturbed by the failure of plate-tectonic theory to explain why slabs descend on an inclined plane rather than vertically or like a coin in honey, like any heavy body.[89]

Laboratory studies of certain rocks indicate that they have been subjected to "ultra-high pressure" (UHP). This is because they have undergone "ultra-deep" subduction, down to depths exceeding 150 kilometers, and have later been exhumed. Numerous geological episodes have produced UHP rocks, but the first and most predominant[90] of these events did not occur until about 540 million years ago.[91] Studies have linked the production of the oldest UHP rocks to an abrupt increase in the Earth's rate of heat loss beginning just before 540 Ma ago,[92] closely coincident with the formation of the first deposits of crystalline colored gemstones (Chapter 1), and with the emergence of the first multicellular animals (Chapter 4).

It has been suggested that the first episode of UHP metamorphism had been caused by an increase in plate thickness[93] due to cooling.[94] Greater plate thickness had caused an increase in the forces pulling slabs downward. I accept these considerations, but just here I prefer to emphasize the emergence of paths of crustal weakness with gentle angles of descent.

To summarize, modern-style plate tectonics involving deep subduction could not have become a geological process until well into our planet's history. Time had been required for slabs to cool. But additional time had been required for multiple erosional cycles. These produced arcs of weakened rock whose widths and angles of descent would allow cool dense slabs to descend, pulled by the force of gravity.

Shallow subduction-like events occurred, or seem to have occurred, here and there early in our planet's history without producing modern-style plate tectonics. But following the first episode of deep subduction, the crust of our planet would have immediately become loosened, thereby facilitating subduction elsewhere along other deeply eroded arcs. This could have led to an abrupt planetwide redistribution of mass[95] and the movement of heavier materials toward the Earth's equator.

The novel feature introduced in this geological reconstruction is the introduction of *an initial condition* of deep, arcuate, 3-D zones of fractured rocks that with time produced downward entrant-ways suitable for the deep subduction of great slabs of dense rock.

Mid-oceanic ridges

When a sufficiently thinned region develops in the Earth's crust, the underlying basaltic material melts by decompression. The basalt does not "boil over" as if it had been heated, but rises gently, following 3-D paths of lowest pressure. Some of the original paths would

* Descending plates are heated on their downward journey, gradually losing their "negative buoyancy." They cease to descend at a depth of about 660 kilometers, apparently due to the abrupt loss of negative buoyancy on encountering minerals whose atoms have been rearranged to form a more dense crystalline structure.

have been along portions of LHB impact scars, but there would be many subsequent geometrical complications, 3-D junctions, and a plethora of shallow linear faults of various origins and ages nearer the surface. Routes of least resistance would be complex, jumping from one arc to another and following linear sections as well. Work by Brian Taylor and colleagues at the University of Hawaii has shown that rifts may jump from one site to another like a zipper that comes open in two or more different places, leading them to conclude that parts of the Earth's crust may be weaker than has been thought.[96]

Repeated separations of one landmass from another might follow nearly identical paths of lessened resistance, as is documented, for example, by the similar traces taken during the opening of the proto-Atlantic ocean and then of the present-day Atlantic itself.[97] Such traces might follow one 3-D craterform and switch to another with a different curvature, whether in the same or opposite sense, following paths of least resistance, arcuate or linear.

Sizes and angles

There is a great deal of variability among present-day subduction zones. The Aleutian arc has a radius of some 1400 kilometers, while the radius of South Sandwich subduction arc in the south Atlantic is only 400 kilometers. The angle of descent of the downgoing slabs is also variable, and there is no evident relationship between the curvature of the arc and the angle of the descent. Although this is generally seen as puzzling, it is consistent with the mechanism proposed here, in which the depth of erosion is a critical consideration.

Mascons

There are regions of the Moon with substantially more mass beneath the surface than expected. Four of the most prominent of these "mascons" (**mas**s **con**centrations) are the Imbrium, Serenitatis, Crisium, and Orientale impact sites, at each of which the Moon's gravity is stronger than average.

The lunar mascons were entirely unexpected when discovered in 1968,[98] and their existence could have easily caused catastrophic failures of the early missions to the Moon. Once they were discovered, the orbits for lunar landers were revised to provide larger target areas and safety margins.

Despite a one-to-one correspondence between large mascons and the circular basins of the lunar maria,[99] the nature of the mascons remained unknown until recently. Specialists had suggested that they are due to kilometers-thick accumulations of mare basalts within the basins, and some amateurs proposed the existence of an enormous iron meteorite below the surface of each mascon, but calculations show that the first idea is insufficient to account for the mascons and the second is not physically possible. In any case, basins should correspond to gravity lows, not highs, so the mascons are actually doubly puzzling. A NASA release in 2006 put it bluntly: "Lunar mascons are a mystery."[100]

The problem was apparently solved in mid-2013 using a number-crunching computational tool called a hydrocode, in which a model of a 3-D target area of the ancient Moon was divided into a large number of small 3-D cells, and each was assigned properties such as density, depth, distance from the center of the impact, temperature, and viscosity. Similar assumptions were made for the impacting body, and computations were carried out for increments of time until movement stopped. It turns out that sufficiently deep lunar impact-excavations will cause dense mantle materials to melt and flow inward, pulling yet other dense materials slowly inward as well. The intuitively expected amount of rebound is prevented by lighter lunar crust that rapidly slides into the hole, forming a rigid cap. A ring with lower gravity is then formed, and ejected material forms a second ring with stronger gravity outside it. No unusual phenomena or assumptions are involved and, after a 45-year wait, we are left with the bland conclusion that mascons are formed by incompleted density adjustments, and that they are "normal."[101]

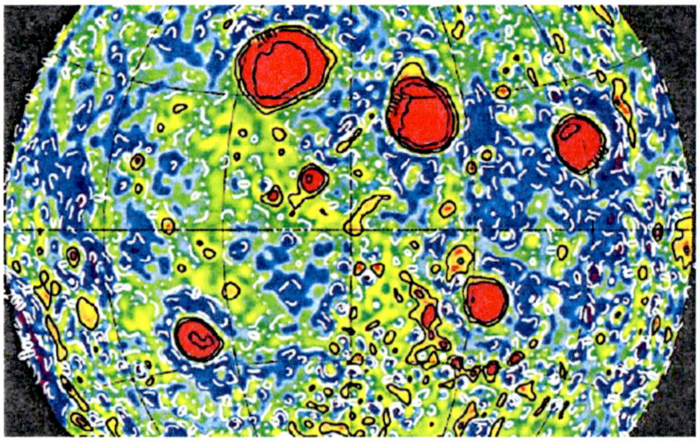

Figure 3.38 Gravitational map of the Moon. The red color indicates the high mass concentration at Mare Imbrium, Mare Serenitatus, Mare Crisium, Mare Humorum, and Mare Nectaris. From Konopliv et al., *Icarus* (2001) vol. 150, fig. 8.

A standard image (**Fig. 3.38**) suggests that the lunar mascons are distributed in a balanced manner, as though their averaged mass were located on the lunar equator. H.J. Melosh had numerically confirmed this impression in the 1970s and then calculated that (1) the axis of the Moon, and (2) the net effect (the "composite spin axis") of the lunar mascons are oriented in nearly the same direction.[102] In other words, a tennis-ball-sized model of the Moon with coins of appropriate mass glued to it in the positions of the mascons would spin in approximately the same orientation as the Moon itself.

We might thus imagine that the mascons, with their associated basalts, had determined the orientation of the Moon's axis. But Mother Nature has not been cooperative in the matter of the mascons, and when Melosh examined the results of his calculations, he found that the mascons could only account for some 20 percent of the mass required to explain the orientation of the Moon's axis,[103] a result subsequently confirmed by others.

Mascons also exist on Mercury and on Mars,[104] where the position of at least one, Argyre, corresponds to a site attributed to a large impact during the Late Heavy Bombardment. So mascons are not unique to the Moon, nor were they difficult to form on certain planetary bodies in the distant past. A review of the recent work by Melosh and coauthors concluded that "It may also be possible to determine whether mascons could have formed in the larger and more active planets, such as Venus and Earth."[105]

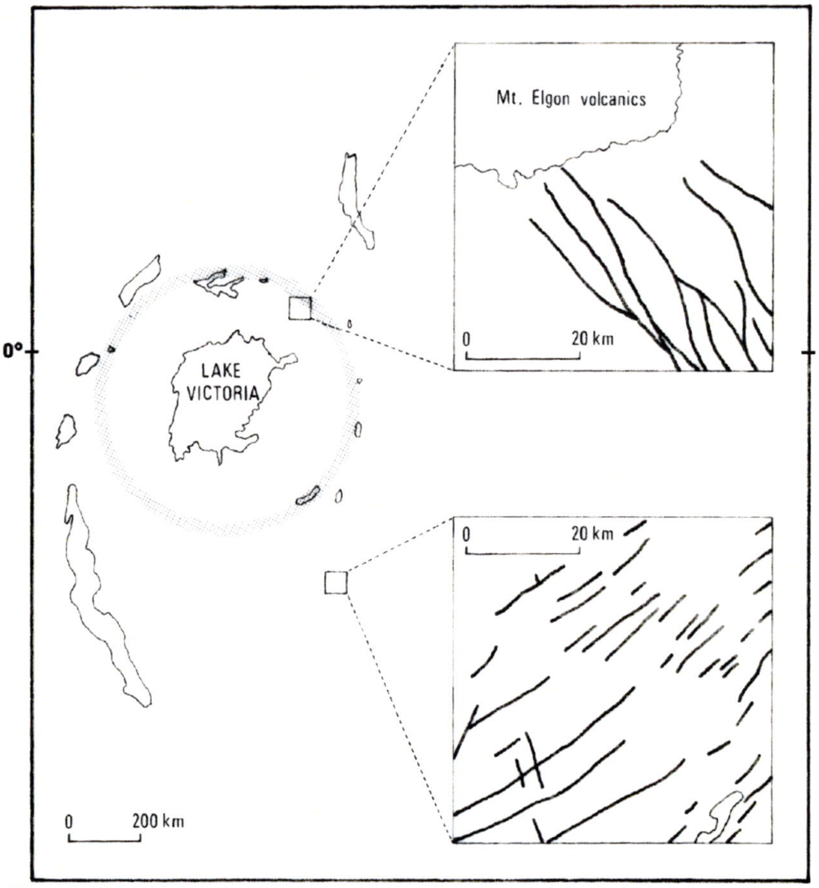

Figure 3.39 Map of Lake Victoria. Insets show the patterns of the geological faults in two selected surrounding areas. The volcanic rocks of Mt. Elgon, which are very young and unfractured, sit atop much older rocks. Map by Janice A. Glaholm.

An obvious question is whether terrestrial mascons, *if they exist*, might be aligned in a similar manner, either along the Earth's Equator or balanced north and south of the Equator. But would such a question make sense on a planet with wandering plates?

Lake Victoria and its neighboring structures to the east and west

Lake Victoria. Former colleague Janice A. Glaholm and I have argued that Lake Victoria — which straddles the Equator — marks the center of an impact scar attributable to the LHB (**Figs. 3.39** and **3.41**).[106] The idea was originally inspired by the geographical arrangement of rifts, lakes, volcanoes, and hot springs around the lake, but once we had the idea in mind, three additional types of evidence also emerged.

One was the pattern of fractures and faults shown on several of the published geological maps of the region. The geological faults on two such maps,[107] greatly reduced and redrawn for **Figure 3.39**, show a concentric orientation around the lake.

A second type of evidence comes from a distinctive group of ancient sedimentary

Figure 3.40 Lake Eyasi at sunset, February 2013. The western wall of the Great Rift Valley can be seen in the distance.

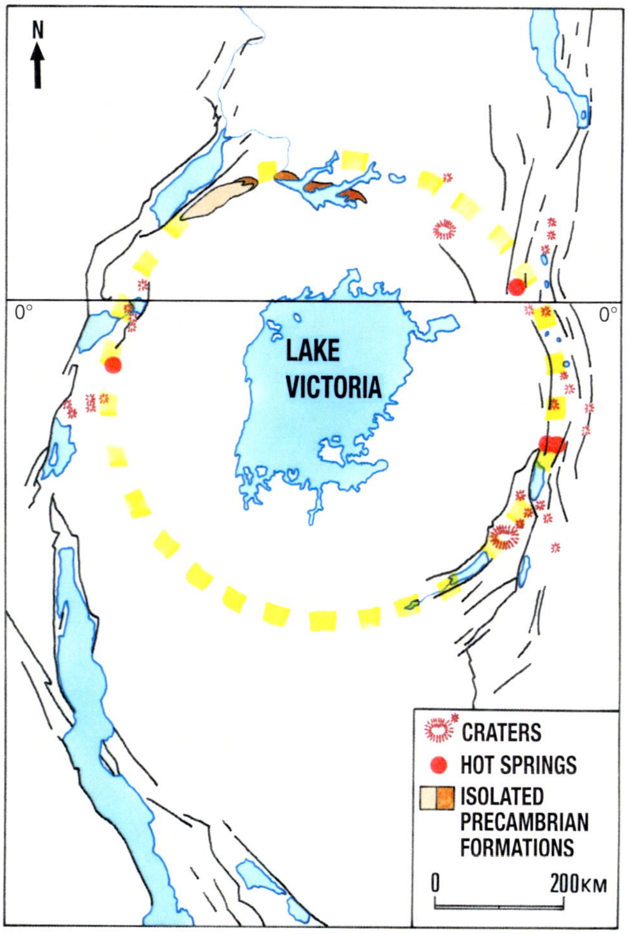

Figure 3.41 Lake Victoria and its enclosing circle. The area shown in pink corresponds to rocks of the Bunyoro Series. The area in red-brown represents rocks of the Kyoga Series, these days grouped with those of the Bunyoro Series by some authorities. Map by Janice A. Glaholm.

rocks[108] known as the Bunyoro Series located northwest of Lake Victoria. These were deposited in a curved depression concentrically disposed around the lake (**Fig. 3.41**). The characteristics of the Bunyoro rocks themselves, their position with respect to the lake, and the arclike shape of their outcrop area are consistent with an origin as the "moat fill," deposited in a moatlike low area between the main rim of the Lake Victoria scar and a hypothetical outer rim.

On the floor of the Great Rift Valley, diagonally across Lake Victoria from the Bunyoro Series, lies Lake Eyasi, a shallow soda lake with no outflow exit. Lake Eyasi (**Fig. 3.40**) occupies a similarly curved depression in an area that has been geologically active over an especially long period of time. At Eyasi, vertical sheetlike intrusions of igneous rocks (dikes) have been emplaced along faults on at least three occasions dated around 2500, 1900, and 900 million years ago,[109] and these dikes are oriented northeast–southwest, tangentially to the Lake Victoria circle.

These and lesser features with depressed topography suggest the existence of a moatlike structure some 20 to 25 kilometers in width surrounding Lake Victoria (**Fig. 3.41**), possibly indicating the long-ago existence of a multi-rimmed scar such as those known from the Moon.

Lake Victoria itself lies in a great depression, and, as noted by Joseph Thomson (1858–1895), one of the early European explorers of East Africa,* Victoria is "unlike most other [large] African lakes" — a reference to the elongate lakes of East Africa — as it "is not bounded by ranges of mountains. The ground descends gradually to its shores; and peacefully the water laps the muddy and marshy beach…."[110] In short, Lake Victoria is not a rift-valley-type lake; it is a great Equator-straddling puddle encircled by one or more rims, perhaps by a moat, and with a concentric pattern of fractures.

The Congo Basin. Immediately to the west of the Lake Victoria circle and also straddling the Equator lies the Congo Basin, which a NASA site characterizes as "a vast, shallow depression which rises to form an almost circular rim of highlands."[111] G.G. Kochemasov, a Soviet geologist who lived and worked in central Africa, compared the basin to "the circular lunar maria of impact origin" and as a title for one of his several communications

* In contrast to other Europeans involved in the scramble for Africa, Thomson was notable for avoiding confrontations with the peoples he encountered. He never killed anyone, nor did he lose any of his men to violence. His motto is said to have been "He who goes gently, goes safely; he who goes safely, goes far." Thomson's gazelle, *Eudorcas thomsonii,* is named after him.

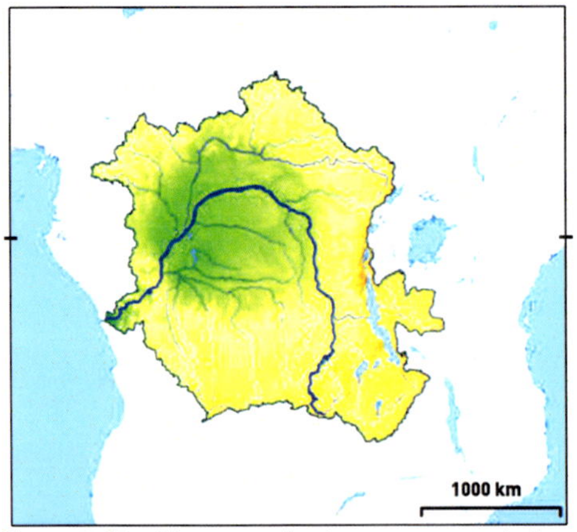

Figure 3.42 The green circular area represents the center of the drainage basin of the Congo River. *The diameter of the "Congo circle" itself is substantially larger* and may extend from the Western Rift Valley almost to the Atlantic.
Source: Imagico: http://en.wikipedia.org/wiki/File:CongoLualaba_watershed_topo.png

to the U.S. Lunar and Planetary Institute, Kochemasov asked, "The Congo Craton: An Old Impact Structure?"[112] His arguments, which are built up of many bits and pieces of local geology, appear to be coherent and at one point he emphasized the "unusual stability and rigidity" of the rocks underlying the Congo Basin.[113] Kochemasov's "circular rim" itself could not be identified sufficiently well to be mapped, but the position and topography of the drainage basin of the Congo River (**Fig. 3.42**), suggest that the Congo Circle may extend from Lake Victoria almost to the Atlantic.

The South American craton. Straddling the Equator opposite the Congo Basin on the other side of the Atlantic is the South American craton (**Fig. 3.17**), understood to include the small slices of ancient Precambrian rocks in the northern Andes. As noted earlier, "when information from drillholes is included, and the younger rocks of the Amazon Basin are ignored, the Guyanan and Brazilian Shields together form a nearly perfect circle."[114] (The Guyanan and Brazilian Shields appear in the northeastern and southeastern quadrants of the circle in **Figure 3.17** but I cannot say whether the "nearly perfect circle" in the citation is identical to the circle in **Figure 3.17** for, as with the Congo, it has not been possible to unambiguously transfer an expert's words onto a map.)[115]

The Indonesian arc. To the west, on the far side of the Pacific, lies the great earthquake-prone Indonesian arc, also situated astride the Equator (**Fig. 3.43** and **Fig. 3.27**). Although it is primarily defined by oceanic crust, the Indonesian structure also cuts into continental rocks in Burma, extending to the edge of the larger Himalayan Circle (**Fig. 3.27**), near the Tengchong volcanic area of China's Yunnan Province. In contrast to the Congo Basin and the South American circles, the Indonesian arc is extremely easy to see on maps of topography, earthquake epicenters, occurrences of oil and gas, or volcanism (**Fig. 3.43**).

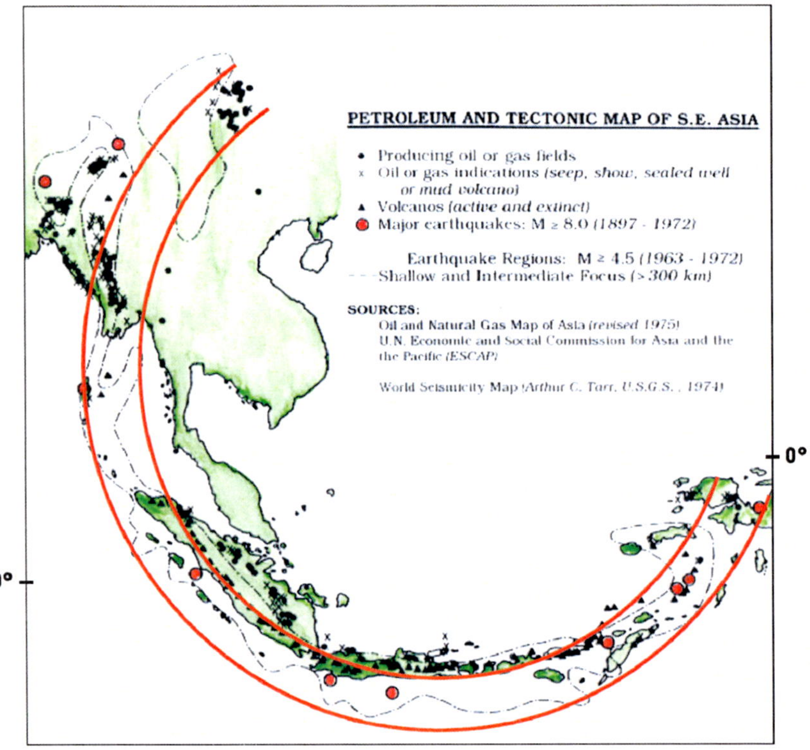

Figure 3.43 Sketch map of the Indonesian Arc, showing the locations of oil and gas fields, volcanoes, and earthquake centers. Modified from p. 96 of Thomas Gold and Steven Soter: "Abiogenic methane and the origin of petroleum," *Energy Exploration & Exploitation* (1982) vol. 1, no. 2, pp. 89–104.

Each of the four circular or arcuate scars discussed in this section straddles the Earth's Equator, three in continental crust and one mostly in oceanic crust, and comparison with the lunar mascons does not seem out of the question. *Something* is concentrated along the Equator.

A personal observation. I propose that a fifth part-circle exists in the area between the Lake Victoria scar and the Indian Ocean, incorporating the only substantial landmass along the Equator not included within one of the four other scars. Although not visible as a continuous feature, this arc is easy to plot on a map because it passes through the three most impressive volcanic centers in East Africa, Mount Kenya (which lies astride the Equator), Tanzania's Mount Kilimanjaro, and the Mount Marsabit volcanic complex of northern Kenya (**Fig. 3.44**).[116] When all three are connected by a smooth arc with circular curvature, it becomes clear that the arc also follows the gently curved one-walled Pangani Rift (**Fig. 3.45**), which extends southeastward of Mount Kilimanjaro.

I have frequently driven along the Pangani Rift, and have even done some prospecting in the region, so on seeing this, I was personally satisfied that I had found the Earth's "last remaining" equatorial scar, and dubbed it the "Pangani Arc." But I was aware of a need for additional evidence to persuade others who have not driven along Tanzania's #B1 highway and do not have a pile of maps of East Africa on hand.

In the event, I failed to find other indications along the actual arc itself, but 20 or 30 kilometers to the east, inside the arc, there are two out-of-place flyspecks on the geological map of Kenya, two unspectacular units of lightly metamorphosed sedimentary rocks, no more than 10 kilometers across in their largest dimensions; see **Figure 3.44**. These are the "Embu Series"[117] (located southeast of Mount Kenya) and the "Ablun Series"[118] (far to the north, within a short distance of the border with Somalia), two minor geological curiosities separated from one another by about 450 kilometers. Although not composed of strictly identical types of rocks, they were astutely compared with one another by the former Chief Geologist of Kenya's Geological Survey, who suggested that an explanation for one might also apply to the other.[119]

The rocks composing the Embu and the Ablun series are themselves unexceptional, but the two are out of place in the context of Kenya's geology, as though olives had been used to decorate a birthday cake. Ignoring them would be to ignore the geological dictum, cited earlier, that "No rock got where it is by accident." With this in mind, I tentatively interpret the rocks of the Embu and Ablun series as sediments accumulated just inside the rim of the craterform scar of the Pangani Arc. If so, they would not be the original LHB crater-fill. Instead, they would belong to a later generation of sediments or slump-rocks within an inherited structure.

If my interpretation is valid, the Pangani Arc would be indicated by six features: from south to north, the curve of the Pangani Rift, Mount Kilimanjaro, the Embu Series, Mount Kenya, the volcanic centers at Marsabit, and the Ablun Series, with a group of rocks associated with Ol Donyo Sabuk mountain as a possible seventh feature (**Fig. 3.44**).

The Equator

Five great circular structures (including the Pangani part-circle) are aligned along the Earth's Equator, each either abutting its neighbor or separated by young oceanic crust. This is not a random pattern. Yet as dictated by the history of our planet and by plate-tectonic theory, the alignment cannot possibly have been established at the outset of our planet's geological history, at the end of the LHB. Instead, these great 3-D circular structures — or the plates on which they resided — must have been individually caused to migrate into their present positions. They must have begun their migrations from a previous geographical configuration that

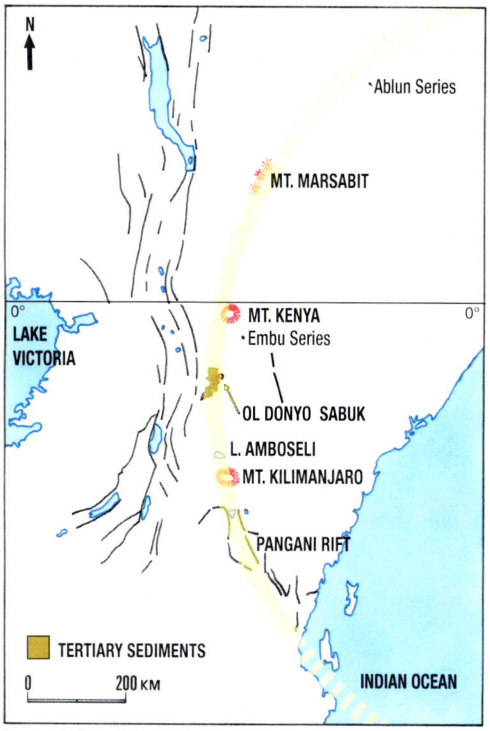

Figure 3.44 Distinctive geologic features define the proposed Pangani Arc, indicated in yellow. Base map by Janice A. Glaholm. See text.

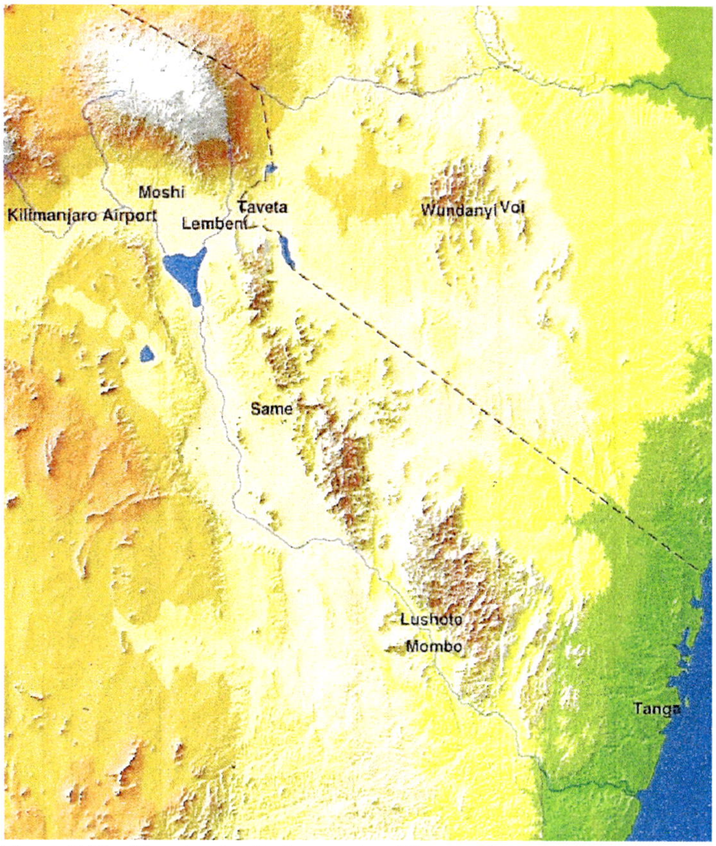

Figure 3.45 Map showing the topography of the smoothly curved one-walled Pangani Rift, which extends southeastward of Mount Kilimanjaro. Digital elevation map courtesy of Christoph Hauzenberger.

had become unstable, which is to say, a configuration that had caused the Earth to wobble on its axis.[120]

Plate-tectonic theory deals with tectonic plates and their movements and migrations. Four of the units currently situated along the Equator are circular and one is a part-circle. Each apparently migrated to its present equatorial position separately. From this we may deduce that many of the larger tectonic plates had been circular units (in 2-D), individually liberated from the remainder of the planet's crust by 3-D fracturing around a 3-D craterform unit.

One of the equatorial scars described just above is the Indonesian Circle, whose interior has the world's highest gravity.[121] This is due to the great quantities of cold, dense oceanic rocks in the underlying crust and mantle, materials accumulated over relatively recent times by subduction events, large and small, such as the great Indian Ocean earthquake of 26 December 2004.

We might imagine, perhaps correctly, that centrifugal force had caused this great excess mass to migrate to the Earth's Equator, thereby reducing the Earth's wobble. (If spun, a tennis ball with one or two coins glued to it will at first wobble and then spin more smoothly with the coins situated on the ball's "equator.") Yet the four other equatorial scars, one South American and three African, involve low-density continental crust and, aside from the region of the Andes, they are *not* characterized by coinlike gravity highs. The basins of the Amazon and the Congo rivers are actually gravity lows.

This confusing situation can be resolved by an ingenious proposal made by Alexander Yakubchuk in 2008,[122] a proposal entirely independent of my own work, that makes no mention of the Late Heavy Bombardment or of circular features.

Yakubchuk contends that supercontinents, at times when they existed, "were always symmetrically centered near the Equator." There they remain for long periods during which their great thickness and size cause them to trap heat rising from below.[123] As they heat up, they become progressively less dense, whereupon the Earth — whose greatest concentration of mass is then no longer situated along the Equator — becomes "spin-unstable." The supercontinents then break apart and the fragments are driven toward the poles[124] with, I would interject, cratons and some other 3-D craterform bodies remaining as intact units, as indeed appears to be the case presented in Yakubchuk's broad-brush schematic maps.

Yakubchuk's reconstructions of the positions of the continents show that in times when the supercontinents were disassembled, "substantially more than 50 percent of the continental masses" were situated in either the northern or the southern hemispheres.

The Earth's spin stability is restored as dense oceanic material comes to occupy the equatorial regions,[125] eliminating the Earth's wobble in the same manner (which Yakubchuk terms "gyroscopic rebalancing") as the spin of the tennis ball stabilizes as the coin moves to the ball's "equator."

The dense oceanic material then in the region of the Earth's Equator is thin and warms rapidly. As it warms, it becomes less dense and the Earth again becomes spin unstable. Continental fragments (too small to have trapped much heat) that had cooled in the meantime then return to the Equator, where they coalesce, though not in the identical configuration as in the previous supercontinent cycle. Yakubchuk buttresses his argument with his insistence, for which he supplies dates, that the plates all start and stop their motions at essentially the same times, showing "remarkable synchronism."[126]

Some alternative reconstructions for older supercontinents do not conform to those of Yakubchuk. But Rodinia, the supercontinent that preceded Gondwana, was indeed centered on the Equator around 750 Ma.[127] Yakubchuk's proposition may be fully valid only from the time of Rodinia's move to the Equator and the initiation of deep subduction, the Earth's crust having been "too tight" in earlier times.

Yakubchuk's scenario is consistent with what is known and suspected of the Earth's geological history for the most recent 750 million years or more: an alternation of supercontinents near the Equator, and dispersed continental fragments scattered between Equator and poles. *The present-day disposition of the continents happens to be intermediate* rather than either supercontinental or fragmented and dispersed.[128]

Extraordinary contentions?

Plate-tectonic movements appear to be driven by gravity, by the cooling of the Earth, and by centrifugal force. Massive 3-D units are driven toward the Equator at all times. Many or all of these units are coherent craterforms dating to the Late Heavy Bombardment.

No continental configuration can be permanently stable because the Earth is a thermally active planet with a "loose" crust on which mass is constantly redistributed. These days subduction, driven by the force of gravity, is the most prominent manner by which mass is redistributed.

Some of my contentions run counter to current scientific convention, but they are not extraordinary. The Late Heavy Bombardment is no longer an extraordinary concept. Gravity and centrifugal force are not extraordinary forces. There is nothing extraordinary about the idea that fractures follow paths of least resistance or the fact that the Earth loses heat into space. My ideas do not require extraordinary demonstrations. The burden of proof is on those who have found it necessary to devise convolute technical vocabulary in discussing variations on the general theme of plate tectonics but who, by ignoring the early history and initial conditions of our planet, have offered incomplete explanations how, when, or why plate tectonics came into being.

Notes and references - *Chapter 3: Early Impact Scars*

1 Saul (1978).
2 In Iceland, "mid-oceanic" eruptions occur on land.
3 Davies (1992).
4 See Stern (2007). Mahadevan, Bendick and Liang (2010) deduce that subduction of oceanic lithosphere can be initiated "under two very different circumstances: foundering of old, cold, thick lithosphere whose negative buoyancy overcomes plate stiffness, or young, hot, thin lithosphere whose stiffness is insufficient to support an applied load."
5 Lowman (1976), who had said much the same thing on earlier occasions.
6 Isachsen (1978).
7 *Target Earth* (1953) by Kelly and Dachille, *Bombarded Earth* (1964) by Gallant, and "The Impact Theory: Asteroids and the Earth-Moon System" (1968) by Cohenour and Sharp were among the more interesting early efforts.
8 http://www.passc.net/EarthImpactDatabase/index.html
9 See Dietz (1961).
10 Grieve (1991).
11 Ferrière (2008): Microphotograph of a quartz grain with two PDF sets. Meta-graywacke sample from the Bosumtwi crater, Ghana.
12 Ferrière et al. (2009): Microphotograph (in plane-polarized light) of elongate ovoid (crescent) to roundish alpha-cristobalite ballen in suevite from the Bosumtwi crater (Ghana).
13 Ferrière (2008): Microphotograph (in plane-polarized light) of a toasted quartz grain with two PDF sets; quartzite clast in suevite from the Bosumtwi crater (Ghana).
14 Ferrière and Osinski (2010): Shatter cone, macrophotograph of horsetailing shatter cone surfaces (in limestone; Steinheim structure, Germany).
15 Garde et al. (2012).
16 For these globally distributed sediments, see Johnson and Melosh (2012); Bottke et al. (2012).
17 Holland (2002).
18 El Albani et al. (2010). Fossils of possible microbial mats have been dated 3500 to 2900 Ma.
19 Adolf Seilacher sees these as pyrite aggregations of nonorganic origin: http://www.arn.org/blogs/index.php/literature/2010/07/02/macroscopic_life_in_the_palaeoproterozoi (accessed 13 March 2013). Others see them as fossilized remains of bacterial colonies.
20 Brain et al. (2012).
21 Shields-Zhou and Och (2011); Och and Shields-Zhou (2012). Campbell and Squire (2010) "suggest that the Gondwanan supermountains were higher than those produced during the assembly of earlier supercontinents and that rapid erosion of these mountains released a large flux of essential nutrients, including Fe and P, into the rivers and oceans, which triggered an explosion of algae and cyanobacteria. This, in turn, produced a marked increase in the production rate of photosynthetic O_2. Rapid sedimentation during this period promoted high rates of burial of biogenic pyrite and organic matter generated during photosynthesis so that they could not back react with O_2, leading to a sustained increase in atmospheric O_2."
22 Chen et al. (2004).
23 Huldtgren et al. (2011). For a discussion, see Lipps et al. (2012).
24 Maloof et al. (2010).
25 Seilacher (1983, 1989, 1992).
26 **Figure 3.8** is from Shuhai Xiao and Marc Laflamme (2009) who provide the following legend: Disparate bodyplans and unique morphologies of the Ediacara biota. (a) *Charniodiscus* frond with a circular holdfast and a large petalodium leaf. (b) *Rangea* displaying fractal, repetitive primary branches and rangeomorph frondlets. (c) Incompletely preserved *Charnia* frond with rectangular modular units within which rangeomorph frondlets reside. (d) *Swartpuntia* frond. (e) *Kimberella* (white arrow) with *Radulichnus* grazing traces (black arrow). (f) *Parvancorina* with bilateral symmetry and anterior–posterior differentiation. (g) *Dickinsonia* displaying shrinkage marks possibly due to muscle contractions. (h) A series of three *Dickinsonia* resting traces (1 = oldest; 3 = youngest) presumed to have been made by one individual. (i) *Spriggina* with bilateral symmetry, anterior–posterior differentiation and possible segmentation. (j) *Tribrachidium* with

triradial symmetry. (k) *Arkarua* with pentaradial symmetry. Scale bars represent 1 cm. Photos (a,e,f,g,h) are provided by J. Gehling, and (b,d) by G. Narbonne.

Specimens identified as *Spriggina*, seen elsewhere, were clearly not bilaterally symmetrical. Instead, they had opposing segments shifted by a half segment-interval.

27 James Hutton (1788), paper presented at the Royal Society of Edinburgh.

28 Rodney Gomes, Hal F. Levison, Alessandro Morbidelli and Kleomenis Tsiganis comprised the original Nice group.

29 Johnson and Melosh (2012), and Bottke et al. (2012). Also see H. Thompson (2012).

30 The type of meteorites in question are the carbonaceous chondrites. The quote comes from Robert (2001). The Earth's nitrogen and other volatiles may have a similar source (Marty, 2012).

31 Head et al. (2010).

32 Cited in *Science* (15 June 2012) p. 1366.

33 Lowman (2002) p. 191.

34 Head et al. (2010).

35 Ryder et al. (2000).

36 Shirley and Fairbridge (1997) p. 196.

37 A dozen or more geological layers dated between 3470 and 1700 Ma contain spherules apparently formed from the condensation of vaporized rock. These "spherule beds," believed associated with the formation of craters up to 40 kilometers in diameter, may be vestiges of tail-end members of the LHB, of LHB impactors whose orbits had been perturbated, or of unrelated impacts. Corresponding craters have not been identified.

38 Some little-altered meteorites contain microscopic grains older than the Sun.

39 Cameron and Benz (1991); S.R. Taylor (1993).

40 Hartmann (2010).

41 A circular scar on the surface of the Earth or any other generally spherical body may or may not appear as a circle when displayed on a flat sheet of paper, depending on the map projection. In general "conformal maps" should be used. I thank Peter H. Dana for the following information:

A map that is "conformal" is a map that has the attribute of keeping local shapes correct with respect to the surface of the earth. A conformal map (like most any map) has scale differences everywhere; in a conformal map, the scale at any point is the same in every direction (for a short distance—very short). What this means is that a circle on a conformal map is the closest to a circle on the ground that one can get for many circles on a single map. The Stereographic, Lambert Conformal Conic, Transverse Mercator, Equidistant Conic, are examples. Of these the Lambert Conformal Conic is best at larger longitudinal distances. The Transverse Mercator maps regions longer in north–south extent, and the Stereographic quite good for regions that are circular or square.

An azimuthal projection, centered at the crater center (only one crater per map), is conformal for that crater circle and no other.

The problem with the conformal map is that conformality is not a global attribute for the map. The further away from a point, the more that particular scale is different.

Another issue is that all these except the Oblique Stereographic have problems at high latitude....

So in general pick conformal projections, avoid equal-area projections, and try to map small areas on the earth.

42 According to the *Milwaukee Sentinel* (10 November 1977), the Rev. David O. Van Slyke traveled the circle on horseback throughout three decades, announcing in 1886 that it was "the veritable Garden of Eden, a place that answers the Bible description of that notable spot better than anything yet discovered." What Donna Stetz, then a graduate student in geology, called the crater rim, Van Slyke called the garden wall.

43 Kerr (1994).

44 Extrapolated from results of laboratory experiments by Donald E. Gault and coworkers.

45 Melosh (1982).

46 Melosh (1982).

47 Schultz and Schultz (1980).

48 Melosh (1982) p. 372.

49 Melosh (2000); Grieve and Cintala (1992).
50 Melosh (2000).
51 Ryder et al. (2000) p. 481.
52 The word *gneiss* (pronounced "nice") is presumed to be of German origin, but its original meaning is not known. The banding is known as compositional banding.
53 Wikipedia contributors, "Craton," *Wikipedia, The Free Encyclopedia,* http://en.wikipedia.org/w/index.php?title=Craton&oldid=579031235 (accessed 20 Dec. 2010).
54 Hoffman (1988).
55 Dalziel (1992), with the archons as perhaps the most dense parts. According to A.J.A. "Bram" Janse, diamond-bearing kimberlites (p. 48) are concentrated in the archons (Levinson et al. 1992).
56 Keith Bloomfield, personal communication, 1980.
57 The Geological Museum is now part of the Natural History Museum. The geological globe no longer exists.
58 Reported ages up to 4280 Ma from the Nuvvuagittuq belt, northern Quebec, have been contested.
59 Mallick and Hodge (1981), who also report anomalous thermal conditions; Miller and Lakatos (1983).
60 Also note the position of the Devil's Tower and the bend in the Big Horn Mountains.
61 Bendick and Bilham (2001).
62 http://books.google.fr/books?id=970cAQAAMAAJ&q=balagan-tas+volcano&dq=balagan-tas+volcano&hl=en&sa=X&ei=1MS2UamCIq2Q0QWk84HgBA&ved=0CDkQ6AEwAQ (accessed 11 June 2013).
63 Whitaker (1981).
64 See Melosh (1982).
65 Scott et al. (2010).
66 Scott et al. (2010). Note also the extreme raggedness of the coast of Asia southeast and opposite Novaya Zemlya.
67 The name of the locality is also rendered Botogol and Batagol. Sources here include the following:
- Touret (2010)
- http://cat.inist.fr/?aModele=afficheN&cpsidt=10523809 [for the occurrence of native silicon]
- http://alkaline.web.ru/2010/Abs/Yatsenko.htm
  http://goldschmidt.info/2013/abstracts/originalPDFs/5739.pdf
- http://proceedings.spiedigitallibrary.org/proceeding.aspx?articleid=995005
- http://www.agu.org/WPS/rgg/45/45.02/articles/S05FEB04.PDF (all retrieved 7 June 2013)
- http:link.springer.content/pdf/10.11.34/S1028334X07010023

Permafrost inside the abandoned Batagal graphite mine.
Photo: Mariya Solovyeva, 2008.

The abandoned Batagal graphite mine (52°21′ N, 100°45′ E), tailings pile, and tracks going into a mine entry.
Photo: Mariya Solovyeva, 2008.

View from the Batagal graphite mine. The hilltop structure in the distance is a hippodrome built for Buryat workers by the French explorer and prospector Jean-Pierre Alibert (1820–1905). Alibert discovered the deposit and worked it for a decade from 1847 to about 1857. He is still considered as a hero by many Buryats, a Mongol people.
Photo: Mariya Solovyeva, 2008.

Batagal, a geological contact photographed by Mariya Solovyeva, 2008. The name Batagal (Батагол) is also transcribed Batagol and as Botogol. The community of Batagal, a short distance to the west-southwest of the abandoned mine, was uninhabited and in ruins at the time of Solovyeva's visit.

68 A.G. Jones (1987); Saul (1978).

69 Jerome and Cook (1967).

70 Godwin (1973).

71 Mardirosian (1973). Individual mines are numbered following Mardirosian (1973).

72 Or, more precisely, lacking economically mineable deposits in which deposition of the ore involved the circulation of fluids, as compiled by Mardirosian (1973).

73 Saul (1978).

74 Saul (1978).

75 Mardirosian (1973).

76 A working definition for the Transition Zone, from Nina Fitzgerald, has it as a 100-kilometer-wide region in central Utah and southern Arizona with structural and stratigraphic characteristics of both the Basin and Range Province to the west and the Colorado Plateau Province to the east. Its boundaries are the subject of disagreement due to various interpretations and use of different criteria. Essentially, extensional tectonics of the Basin and Range has been superimposed upon the adjacent coeval uplifted blocks of the Colorado Plateau and Middle Rocky Mountains. Block faults, the principal feature of the Basin and Range, extend tens of kilometers into the adjacent provinces, forming a wide zone of transitional tectonics, structure, and physiography. There is also evidence that the planet's rigid outermost shell, the lithosphere, may be particularly thin in the region of the ATZ, yet it may be comparably thin in the Basin and Range Province. http://www.blogger.com/profile/08965179274125274725 (accessed 1 January 2011).

Some also hold "that volcanism and extension have been slowly eating away the Colorado Plateau, and the result of this process is the Transition Zone." Some recent maps of the ATZ show it as smaller than indicated in the past. http://www.fs.usda.gov/detail/r3/landmanagement/resourcemanagement/?cid=fsbdev3_022250 (accessed 21 July 2012).

77 Although the method used here (Saul, 1978) has considerable strengths, it is highly sensitive to the direction and the height of the light source. In consequence, replication of results is not necessarily an easy matter.

The numbering given to mineral deposits or mines by Mardirosian (1973) is as follows:

(56) Banner, Xmas, Troy, Dripping Springs, and Hayden
(58) Bobtail and Samsel
(59) Canyon Creek
(64) Ellison
(66) Globe-Miami [and the Old Dominion group]
(67) Green Valley / Payson
(73) Pinto Creek and Pinto Valley
(75) Richmond Basin, MacMorris, and McMillen
(84) Aravaipa
(93) Stanley, Stanley Butte, and Quartzite Mountain
(106) Cave Creek, Gold Cliff, and Golden Reef
(112) Magazine and Gray's Gulch
(226) Goldfields
(227) Hewitt Canyon
(230) Martinez Canyon
(231) Mineral Creek, Ray and Kelvin
(232) Mineral Hill
(239) Pioneer, Magma, Silver King, and Superior
(241) Riverside, Kearney, and Rare Metals
(242) Saddle Mountain
(302) Squaw Peak

78 Pierre Nicolini (1980) ch. 1, p. 4.

79 Weatherley and Henley (2013).

80 Saul and Glaholm (1981) presented the "Stockholm Circle" as a vestige of the LHB.

81 I have rendered the translator's or author's "megacleavage zones" as "breccia zones."

82 With thanks to Valery Mikhailovich Moralev (1928–2003), as his name was more usually transliterated.

83 Hindley (11 August 1977) p. 353, who again took up the subject three years later (20 March 1980).

84 Silver and Behn (2008); also see Hansen (2007) and the sources she cites.

85 Stern (2007).

86 Stern (2007) p. 578.

87 During continent-to-continent collisions, the difference in densities between the plates is relatively small. Subduction in such cases, if it occurs, may be less pronounced than when an oceanic plate is involved. Walter Balmer (personal communication, Dec. 2013) uses the Alps as an example of this phenomenon. Mantle material might peel off the bottom of an overly buoyant plate that "tried to subduct," with the delaminated mantle material returning to depth but not the overlying rocks.

88 Mantle movements and shearing are slow processes. If outgassing occurred regularly along the same paths, and rejuvenated them on short time scales, matters would be more complicated.

89 Vladimir V. Beloussov, personal communication 1969; also see Beloussov (1979).

90 Michael Brown (2008).

91 Maruyama and Liou (1998), who have the first UHP conditions preceded by a transitional period from 750 to 540 Ma characterized by intermediate conditions.

92 Maruyama and Liou (1998). Heat flows from hot objects to cold, and since the Earth and other planets are relatively hot, their upper layers must lose heat into the cold of space. One way by which a planetary body may lose heat is by massive outflowings of lava. This occurs on Jupiter's satellite, Io, where lava flows produce full resurfacings. On the Earth, great "flood basalt" outflows have also occurred, and are closely associated with mass extinctions (Courtillot and Renne, 2003, and more recent papers by Courtillot with various coauthors). Flood basalts presumably commence in places where the Earth's crust is relatively weak. The figure reproduced on the following page, with the permission of Vincent Courtillot, shows the correlation between mass extinctions and episodes of flood basalt activity.

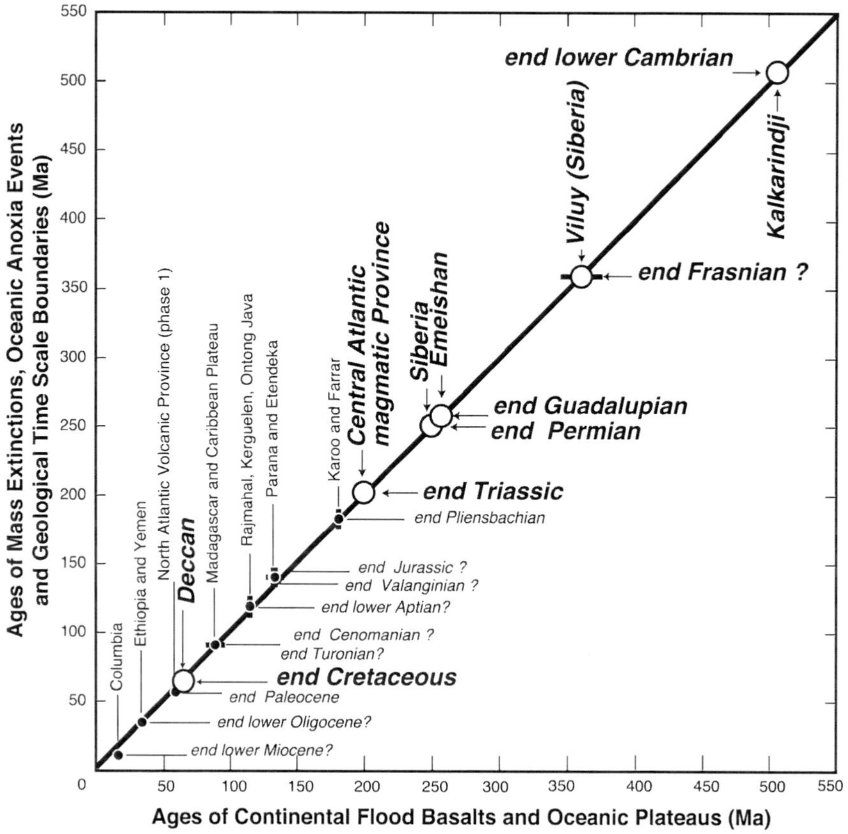

The Columbia River area in America's Pacific Northwest, India's Deccan traps, and the great basaltic areas of Siberia are among the best known terrestrial flood basalts. Such eruptions are probably not melting events as such. Instead, they appear to be drainage events that allow accumulations of previously melted materials to drain upward, reaching the relative cold of the Earth's surface domain through thin rift-like openings (Silver et al., 2006).

93 Maruyama and Liou (1998).

94 Brown (2008).

95 Gold (1955); Kirschvink, Ripperdan and Evans (1997).

96 See Goddard (1995) p. 14.

97 Murphy et al. (2008).

98 Muller and Sjogren (1968).

99 O'Leary (1968) p. 1309.

100 http://science.nasa.gov/headlines/y2006/06nov_loworbit.htm

101 See Melosh et al. (2013); Montesi (2013); and other materials in the same issue of *Science* (28 June 2013). In an interview, Maria T. Zuber, one of Melosh's coauthors, saw "an indication that the Moon's interior may be incredibly fractured" (http://phys.org/news/2013-03-deep-moon-interior-early-history.html#jCp, retrieved 14 September 2013). Also see Wieczorek et al. (2013).

102 Melosh (1975).

103 Melosh (1975).

104 *Science News* (29 April 1978) p. 281.

105 Montesi (2013).

106 Saul and Glaholm (1981).

107 Sanders (1965). Julian et al. (1962). The concentric fault pattern appears on other regional geological maps, but not all.

108 Although it may be older, the Bunyoro Series is usually assigned to the Late Precambrian; see Bjørlykke (1981).

109 McConnell (1974) p. 402.

110 Joseph Thomson, *Proceedings of the Royal Geographical Society* (1884–1885 opening session) p. 707.

111 http://www2.jpl.nasa.gov/srtm/africa_radar_images.htm (accessed 21 November 2010).

112 Kochemasov (1983).

113 Kochemasov (1983), p. 378, generalized in an attempt to explain the "stable nuclei of the continents." Those attempting to elaborate on this will have to track down Kochemasov's numerous publications in English and Russian, some of them quite obscure.

114 Keith Bloomfield, personal communication, 1980.

115 The South American craton and the circle of the Lesser Antilles are separated by a tangential feature.

116 All three of these volcanic centers are underlain by Precambrian rocks with Pan-African ages. Mount Elgon, an extinct shield volcano on the border of Kenya and Uganda, also straddles the Equator.

117 Bear (1952).

118 Thompson and Dodson (1958), which reports work carried out years earlier.

119 William Pulfrey's foreward to Thompson and Dodson (1958); and Kenya Mines and Geological Department, *Annual Report 1953*, p. 9; also see Bear (1952) and the main text of Thompson and Dodson (1958).

120 In the vocabulary of physics, the Earth's maximum moment of inertia became decoupled from its axis of rotation. In everyday language, the Earth began to wobble.

121 http://www.csr.utexas.edu/grace/gallery/gravity/

122 Yakubchuk (2008); also see Gold (1955); and Kirschvink et al. (1997).

123 Yakubchuk (2008).

124 Yakubchuk (2008). Excess mass is pushed toward the equator when a quasi-rigid rotating body aligns its principal moment of inertia with its spin-axis; see Gold (1955). The Earth's crust might shift as a unit or piece by piece in relation to the mantle.

125 Yakubchuk (2008).

126 Yakubchuk (2008) p. 387. Yakubchuk suggests "that the Earth can best be approximated as a constantly rebalancing gyroscope" (p. 396).

127 See Kirschvink et al. (1997), and Z.X. Li et al. (2004).

128 More generally, the existence of non-equilibrium situations — the movement of tectonic plates in this case — commonly indicates that the subject matter is time-factored, i.e., that history is at play.

4

Animals and Cancer

Nothing in biology makes sense except in the light of evolution.
—Theodosius Dobzhansky*

LEONARDO DA VINCI held that "a painter should begin every canvas with a wash of black, because all things in nature are dark except where exposed by the light." I am no artist, but this is one of my favorite sayings. Another is Leonardo's urging to "realize that everything connects to everything else," echoed by thinkers as different as John Muir and Richard Feynman.

Earlier chapters treated the following topics:

- the formation of transparent gemstones, many of which are accompanied by an organic-like odor,
- the presence of methane and petroleum of nonbiogenic origin deep in the Earth, and
- the 4 billion-year-old impact scars that traverse the entire crust of our planet.

All three come together in southeastern Kenya and northeastern Tanzania in the region between Mount Kilimanjaro and the Indian Ocean. Hundreds of occurrences of crystalline colored gemstones (CCGs, defined in Chapter 1) have been discovered in this area and although most are of trivial economic interest, six are major deposits, together producing hundreds of millions of dollars worth of gems. All six fall on the circumference of a nearly perfect geometric circle, or very nearby on faults emanating from the circle (**Fig. 4.1**). All six are also situated in rocks that emit an "organic" odor when struck or crushed and, in additional geological detail, many of the rocks that host the gems dip directly away from the circumference of the circle.

The large white arrows in the northeastern and southwestern margins of **Figure 4.1**† indicate the angle at which East and West Gondwana collided,[1] forming a great north–south "supermountain" (Chapter 1) in one of the main acts of the "Pan-African Event," 650 to 500 million years ago.

---

* Theodosius Dobzhansky, "Nothing in Biology Makes Sense Except in the Light of Evolution," *American Biology Teacher,* vol. 35 (1973) pp. 125–129.

† Previously presented at the Kazan Workshop on Abiotic Hydrocarbons, April 2013, and at the 33rd International Gemmological Conference, Hanoi, October 2013.

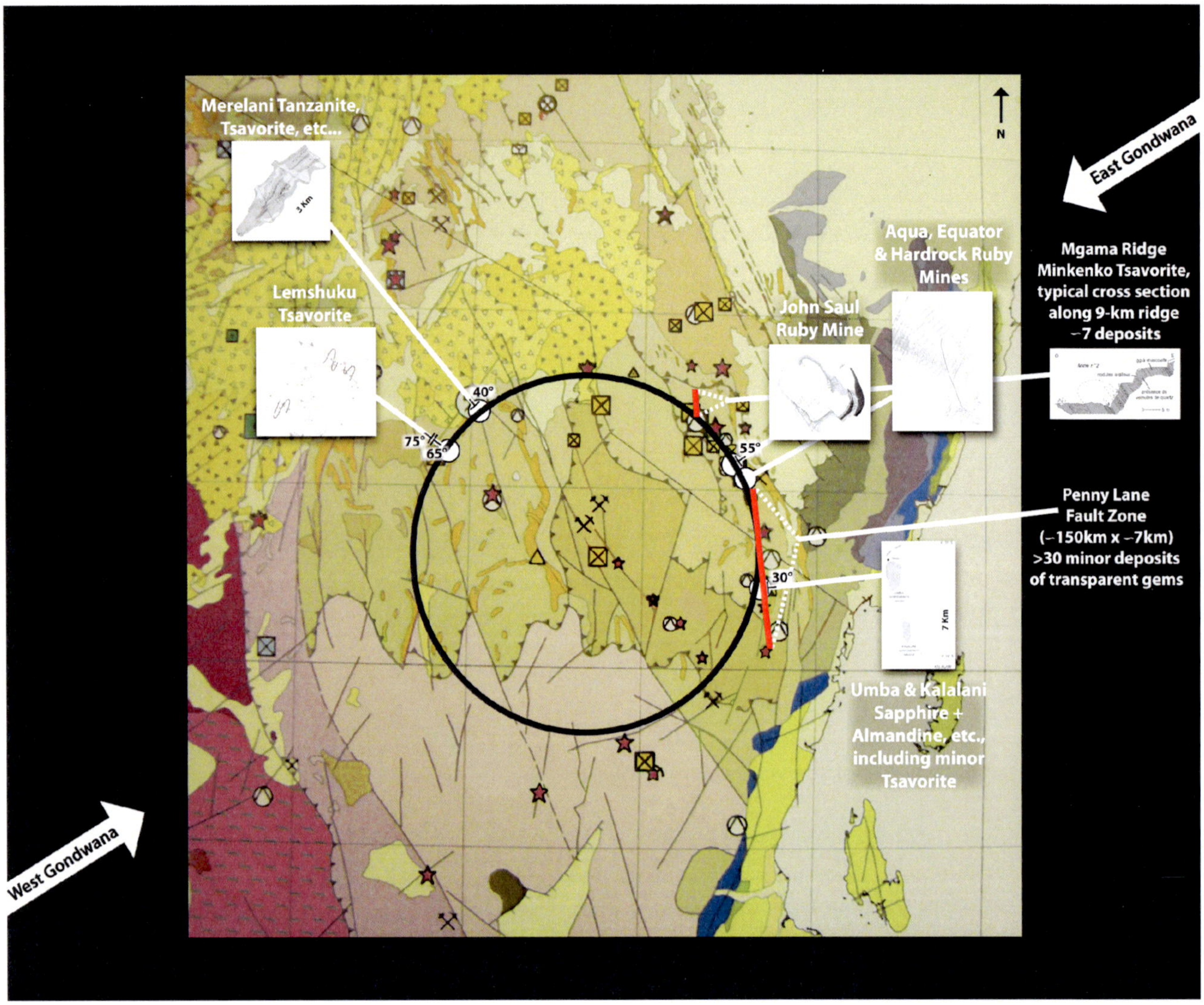

Figure 4.1 Geological map of southeastern Kenya and northeastern Tanzania showing the six main regional deposits of transparent colored gemstones. (Base map: Deschamps et al., 2004). The map shows the disposition of gem deposits in relation to the circumference of a circular scar. The two red lines indicate faults along which multiple occurrences of transparent gemstones are located. The longer fault zone, tangential to the circle, has been dubbed "Penny Lane." For a linear feature tangential to circles in South America, see Chapter 3, note 115.

The "Cambrian Explosion"

Most people interested in the history of our planet identify much of this particular interval of time with the "Cambrian Explosion" (**Table 3.2**), the poorly understood sequence of events during which complex, multicelled animals made their first appearances. With that, the world's oceans became the habitat of creatures that can be recognized as the distant ancestors of all the animals that populate today's seas and lands. In this chapter I shall add two additional matters to my list of "things connected":

- the "Cambrian Explosion" of multicelled animals, and
- cancer, as a pathology that can only exist in the context of multicellularity.

Events in the seas around the times of the Cambrian Explosion (a term I use broadly throughout this chapter) were not limited to the emergence of metazoans.[2] For within this same time span there also appeared marine animals equipped with shells and other hard parts, commencing with protists with their small or microscopic shells, and followed by certain metazoans. ("Hard parts" is a proper scientific term among those who work with fossils.) During these times, with an intriguing exception just before the Cambrian Explosion, the oxygen content of the seas had been generally rising,[3] a matter long understood to have somehow led to the emergence of complex life.[4]

Before the Cambrian Explosion, many or most single-celled creatures, prokaryotes and protists alike, had lived in environments with very little oxygen. Some oxygen — a highly reactive substance — had been present here and there, now and then, in the marine environment, but sometime well before the Explosion, the concentration of oxygen in the seas reached levels that, I propose, would have been toxic to certain single-celled creatures (anaerobes) that had long thrived in oxygen-poor (anaerobic) circumstances.

The ancestors of these single-celled creatures would have been survivors of past encounters with lesser concentrations of oxygen. In consequence, some would have possessed ways to cope, mechanisms that chemically defanged the oxygen and eliminated oxygen-rich molecules. Pure Darwinism would have been involved. Those that happened to be able to excrete oxygen, or to avoid it, would have survived and produced descendant generations; those unable to do so would have died out.

Various traces and indications appear in the fossil record in the run-up to the emergence of multicelled animals; most are controversial and taken one by one, all are difficult to interpret. Some are listed below, along with published dates that are subject to changes, as is normally the case in fixing an exact date on the first appearance of anything. After presenting them, I will avoid plunging into the nitty-gritty of descriptive and technical detail for the same reason I have chosen to define the Cambrian Explosion as broadly as possible, for I intend to treat them all together and leave others to continue to work out the exact sequence of events.

Some definitions

Metazoans, also called ***multicelled animals***, are complex animals composed of specialized cells. Each cell type has a particular role or roles within the animal.

A ***tissue*** is an ensemble of similar cells that together carry out a specific function.

Metazoans are ***eukaryotes*** whose cells, by definition, possess nuclei bounded by membranes, and internal organelles with specialized functions. Complex plants (metaphytes), fungi, and many single-celled creatures are also eukaryotes.

Single-celled eukaryotes such as amoebas are called ***protists***.

Bacteria, whose cells do not have membrane-bounded nuclei or internal organelles, and are generally far smaller than eukaryotic cells, are ***prokaryotes***.

Colonial bacteria are composed of aggregates of non-special cells.

Anaerobes are organisms whose metabolism does not require the presence of free oxygen.

The ***collagen*** family of molecules are the main structural protein in the connective tissues of animals. The word "collagen" comes from the Greek *kolla* meaning glue and *-gen* meaning to be produced.

- Protists balled up as cysts and eukaryotic algae,[5] previously identified as animal embryos[6] (~635 Ma).
- The "Ediacaran biota," a broad variety of sizable creatures lacking clear left–right symmetry, dated approximately 576 to 542 Ma.[7] See **Table 3.2** and **Figure 3.8**. Many were quiltlike. Almost all Ediacarans were soft-bodied, but one Ediacaran resembling a sponge had spicules made of some unknown hard substance.[8] A few specialists speculate that the Ediacarans could have been giant protists.[9]
- Tracks, trails, and pellet traces, collectively known as trace fossils, the oldest of which may date around 555 Ma.[10] The tracks and trails have long been attributed to multi-celled creatures, but during recent research dives to 750–780 meters in the Bahamas, "a multitude of grape-like objects" up to 3 centimeters in diameter, and subsequently identified as amoebas were observed associated with seemingly identical tracks up to a half meter in length.[11]
- Horizontal galleries in soft sand,[12] commonly but not universally[13] interpreted as having been excavated by collagen-poor wormlike creatures with "hydrostatic skeletons" (i.e., with bodies whose rigidity was obtained by maintaining internal body fluids under pressure), dating to perhaps 545 Ma.
- Short linear or zigzag U-shaped burrows in firm clay with distinctively marked lower surfaces,[14] dated around 555 to 542 Ma.

Dates mentioned in this listing fall within the Pan-African, the unique episode discussed in Chapter 1 during which transparent colored gemstones (CCGs) first crystallized. **Figure 4.2** is a timeline on which the dates of the oldest known fossils of nonmicroscopic animals can be compared with the times of gemstone formation.

Figure 4.2 Timeline to scale, showing dates of the earliest non-microscopic animals and periods of formation of crystalline colored gemstones.

- Traces, beginning toward 540 Ma, held to have been made by creatures whose lifestyles had been influenced by the availability and concentration of oxygen.[15]
- Cylindrical chambers open at their tops.[16]
- Worm-tubes lined with flakes of mica[17] from shortly after 542 Ma.
- Protists with sand grains stuck to their surfaces,[18] presumably due to the presence of sticky biomolecules.
- Protists that secreted calcium carbonate.[19]
- Burrows; and a grand variety of fragments ("small shelly fossils"), starting around 542 to 520 Ma, derived from the remains of
  - corals and other metazoans with radial symmetry,
  - mollusks, arthropods, and other recognizable creatures with left–right symmetry, and
  - unidentified creatures, treated by some specialists as pieces of tiny broken-up suits of bio-armor, vestiges of a hypothetical arms race, as suggested by the concurrent appearance of burrows.

The nature and significance of these events are confusing to specialists and nonspecialists alike.

My listing of first appearances of various fossils, which is in roughly chronological order, does not emphasize the one particular development that may help explain all the others. Toward 550 million years ago, there appeared diverse multi-celled creatures equipped with mineralized "worm shells," spicules, external plates, toothlike structures, carapaces, reinforcements, and unclassified hard parts. A phenomenon — *biomineralization* — had suddenly become an important biological activity for many groups of animals, some of which were very different from others. Furthermore, *it appears that each group acquired this ability independently.*[20]

All are familiar with the concepts "Animal Kingdom" and "species." Other levels exist between the two: phylum, class, order, family, and genus, and these may be adorned with modifiers, to give subclass, superfamily, etc. Still other levels, such as grandorder, have also been defined for some groups of animals.

The phylum level is less commonly modified. This is because each of the approximately 35 phyla (plural of phylum) is characterized by a unique "body plan." Familiar representatives of familiar phyla are humans, goldfish and dinosaurs (Phylum Chordata), crabs, mosquitoes and trilobites (Arthropoda), snails, clams and ammonites (Mollusca), flukes and other flatworms (Platyhelminthes), and various types of sponges (Porifera), but many phyla are categories of generally unfamiliar varieties of "marine worms" — arrow worms, round worms, acorn worms, ribbon worms, etc. — each with its own entirely distinct body plan. Animals "between phyla" do not exist and have never existed.

With one apparent exception,[21] the phyla all came into being around the time of the Cambrian Explosion. And very shortly thereafter, creatures from numerous different phyla began to manufacture hard parts, apparently independently. These were constructed using a puzzline array of different substances: aragonite, calcite, phosphate (most commonly as hydroxylapatite), chitin, silica,[22] compact collagen fibers,[23] and agglutinated sand.[24]*

In some biomineralized fossils, the working-ends of "teeth" (by which I mean any hard part that would have been subjected to especial abrasion in order to keep the animal nourished) are reddish or blackish due to reinforcement by concentrations of iron.[25] In similar manner, certain mollusks (chitons) in today's seas have iron oxide (magnetite) capping at the ends of their tongue plates, enabling them to dislodge and eat algae living a few millimeters below the surface.[26]

In another matter, which, as I will try to show, is not unrelated, members of *at least* four phyla (Chordata, Arthropoda, Mollusca, and Platyhelminthes) are known to be subject to cancer.[27] But animals belonging to any two phyla have been separate from one another ever since the Cambrian Explosion. It is possible, I suppose, to argue that incidences of cancer are due to the horizontal transfer of "cancer genes." But I prefer to think of the risk of cancer as *an initial condition* of multicellularity, as a trade-off or price we pay for being complex multicellular animals,[28] with the most complex animals, and those undergoing the most rapid evolutionary changes, tending to be the most susceptible.[29]

I have tried in a series of professional articles[30] to make sense of all this, to combine our highly unsatisfactory understanding of the Cambrian Explosion with our highly unsatisfactory understanding of cancer. My articles have received few citations but, as my late friend Tom Gold used to say, "These things take time." Here I will try again in a less technical style, borrowing heavily from my earlier writings. My reconstruction of events starts a long time before the Cambrian Explosion.

Sticky science

In June 1987, a mishap occurred during the drilling of an oil well on the Swedish meteorite impact site at Siljan (**Fig. 3.35**), a wildcat attempt to find commercial quantities of oil and to demonstrate that petroleum could exist in severely fractured crystalline rocks. The drill became stuck at the very considerable depth of 6 kilometers and when freed 10 days later and hauled up, the lowest 10 meters of drill-pipe were found clogged with an unexpected sticky black material with a particularly nasty odor to it. About 60 kilograms of the material was eventually "recovered," if that is the right word, because most of it was discarded and buried in haste due to its stench. The on-site chemist stated that the odor was of bacterial origin but was puzzled by the black color. A small quantity was preserved in a plastic bag and the color was eventually determined to come from exceedingly fine grains of magnetite.[34] Months later when I had a chance to examine some of the extracted magnetite, it was dry and no longer sticky, but retained a lingering odor. The black material was so fine-grained that it remained in the pores of my fingers, as would jewelers rouge.

Any bacteria living at a depth of 6000 meters would have been anaerobic and intolerant of the oxygen introduced during the drilling operation. The exceedingly fine grains of magnetite, Fe_3O_4 can be interpreted as the product of bacterial efforts to sequester the oxygen, and the organic smell as evidence that their efforts had been generally unsuccessful.

Oxygen and oxygen toxicity

Our record of life starts approximately 3900 to 3500 million years ago in a world in which molecular oxygen, O_2, was very scarce, though not entirely absent. It was in these chemical surroundings that many fundamental and enduring metabolic pathways were established. Around 2900 to 2700 million years ago, colonial cyanobacteria (algae) began to harness sunlight, transforming CO_2 and H_2O into sugars and other carbohydrates by photosynthesis and generating oxygen as a waste product. Oxygen did not appreciably accumulate immediately, however, for it was consumed by iron dissolved in the seas and by reducing gases issuing from undersea volcanoes.[31] A period of fluctuating oxygen conditions most probably ensued, ending toward 2500 Ma with the stabilization of the cratons[32] (Chapter 3).

As the oxygen content of the seas rose, marine creatures, including bacteria, would have at times been exposed to toxic concentrations of highly reactive free oxygen and its derivative molecules and ions. Some perished. Others survived.

Magnetite and magnetotactic navigation

Sometime before 1900 million years ago, at a time when the seas were still rich in dissolved iron and poor in oxygen, certain variant bacteria† acquired the ability to

* Aragonite is $CaCO_3$ with orthorhombic symmetry; calcite, $CaCO_3$ with trigonal symmetry; hydroxylapatite, $Ca_5(PO_4)_3(OH)$; chitin, $(C_8H_{13}O_5N)_n$; silica $(SiO_2{\cdot}nH_2O)$.

† Most bacteria are strictly unicellular and microscopic, but the cyanobacteria, which are individually microscopic, may form large colonies of near-identical nonspecialized cells.

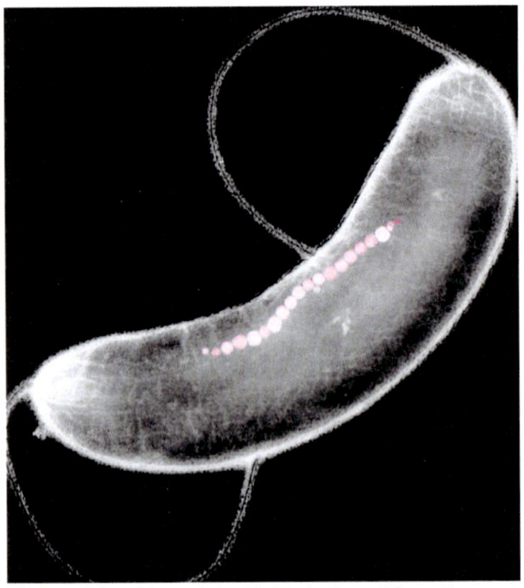

Figure 4.3 Magnetotactic bacterium, *Magnetospirillum magneticum*, from a pond in New Hampshire. The opaque black magnetite crystals have been highlighted in pink. Longest dimension: approximately 0.003 mm. Source: NASA Johnson Space Center.

precipitate the black iron-oxide mineral magnetite, a readily formed substance composed of three atoms of iron and four of oxygen, Fe_3O_4. Descendants of these "anaerobic prokaryotes" survived generation after generation while continuing to precipitate magnetite, an ability that provided them with a detoxification mechanism, not for excess iron, as has been suggested, but *for oxygen*.

In general, I would imagine, the magnetite would have been formed outside the bodies of the bacteria (**sidebar**) but accidents happen, and some bacteria would have burdened themselves with internal crystals of magnetite. Chains of tiny magnetic crystals, all with the same north–south magnetic polarity, are present in some modern bacteria (**Fig. 4.3**).

These days living anaerobic bacteria that incorporate chains of magnetite crystals are commonly found in sharp transition zones where oxygenated water overlies oxygen-free water. There, the twist (torque) imposed on the chains by the Earth's magnetic field — which has a vertical component as well as the north–south horizontal component — causes them to move downward when encountering toxic concentrations of oxygen.[33]

Magnetite grains with comparable characteristics have since been found in salmon, pigeons, and humans and in certain single-celled eukaryotic algae.[35] The magnetite crystals in these eukaryotes are so similar to those in bacteria that "independent invention" appears excluded. This suggests that eukaryotes had acquired the prokaryotic mechanism sometime well before the Cambrian Explosion when a single-celled eukaryote incorporated a magnetite-sequestering bacterium.[36]

The generally underappreciated process during which one cell incorporates another is known as symbiosis. Its importance is emphasized by Nikolaevna Khakhina, a Russian botanist and historian of science, who, with a touch of genius, suggests that if symbiosis is the author of speciation, natural selection is the editor.[37]

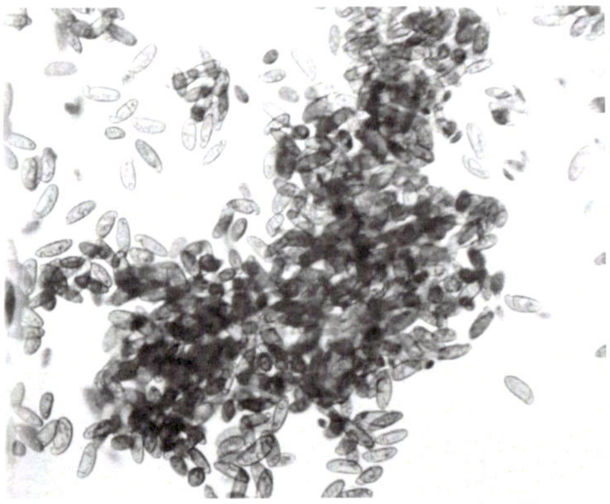

Figure 4.4 "Typical clumping of anaerobic ciliates when placed in oxygenated water. The cells consume oxygen and an anoxic microzone is eventually produced by the clump. The photograph shows a population of one ciliate species (*Plagiopyla frontata*) and each cell has a length of about 0.1 mm" (Fenchel and Finlay [1995] figure 3.4; figure and caption reproduced with permission).

Survival in a toxic environment

Concentrations of oxygen would have killed many unicellular eukaryotes (protists), but some would have survived by navigating downward; by burrowing; by clustering together as do certain oxygen-fleeing protists (ciliates) when placed in oxygenated water (**Fig. 4.4**); or by deploying detoxification mechanisms. Use of one mechanism would not have precluded use of another, and the mechanisms employed could have been usable for neutralizing the effects of other types of toxic molecules as well.

Collagen

Detoxification through the production of magnetite dated to times when the seas were rich in iron.

The amount of iron dissolved in the Earth's oceans fluctuated over time during the first 80 percent of the Earth's history, as evidenced by the cyclic deposition of what are called banded iron formations, but after 600 Ma or thereabouts, iron was no longer readily available from seawater. Detoxification would thereafter have to employ ions other than those of iron, and would produce sequestration products other than molecules of magnetite.

Natural selection would have favored the survival of creatures that used readily available materials and excreted oxygen-rich or oxygen-containing molecules that were inexpensive to shed, with "inexpensive" measured in terms of energy (or molecules of adenosine triphosphate, ATP) per atom of oxygen excreted.

At the time of the Cambrian Explosion, when the first multicelled animals came into being, such oxygen-containing molecules included collagen. We know this because collagen, which is essential for the formation of tissues, is present in all metazoans of all phyla, arrow worms to aardvarks.

Members of the collagen family of molecules have been described as "oxygen expensive."[38] Their characteristics include

- the absolute requirement of molecular oxygen, O_2, for their formation,[39]
- their repetitive structure, with periodic spacings that vary according to the particular variety of collagen,
- accumulation outside the cells (in extracellular space),
- production in great quantities,
- their presence in sponges (which some specialists do not treat as metazoans because sponges do not possess tissues), and
- their absence in living unicellular eukaryotes (protists).[40]

Collagen could not have been formed until the concentration of oxygen in the seas had exceeded some minimal threshold.[41] Yet just before the appearance of metazoans during the events of the Cambrian Explosion, the level of oxygen in the seas was actually falling.[42] So the threshold for the formation of collagen must have been passed well before, perhaps by 576 Ma, by which time the seas were already depleted in iron and when the first "macroscopic" Ediacaran-type fossils appeared.[43] (Collagen could not have been produced earlier than 580 Ma because the deep ocean was then oxygen free.[44])

Yet the complexity of multicelled creatures living in Ediacaran times immediately before the Cambrian Explosion had been constrained, perhaps

- because they respired, ate, and excreted by osmosis and absorption as is suggested by the frondlike and quilted body plans typical of Ediacarans, or
- because the quality of collagen was poor due to lack of sufficient time for adequate winnowing by natural selection,[45] or
- because the mechanism (gastrulation) by which the early embryonic single layer of cells reorganizes itself into ectoderm, mesoderm, and endoderm had not yet emerged,[46] or
- because of inherent limits to the efficiency of animals lacking well-defined left–right symmetry.[47]

But behind any and all such considerations is a grander possibility: that creatures in these times were starved for nutrients, specifically for the phosphorus required in order to form sufficient quantities of the energy-currency molecule ATP to maintain active animals of substantial size.[48]

Phosphorus and the Pan-African

Exceptional quantities of phosphorus appear to have become available around this time. They were derived from the erosion of the Transgondwanan supermountain,[49] the extraordinary topographic feature discussed earlier (Chapter 1), formed by the oblique collision of East and West Gondwana in late Pan-African times.[50] Erosion of this mountain was quantitatively and qualitatively unique, for it occurred in conditions of high rainfall at a time when microscopic soil biota may have already "evolved to the point that they could accelerate chemical weathering"[51] but before the appearance of rooted plants that might retain the soil.[52] The consequence was a uniquely large and rapid flux of phosphorous into the oceans commencing about 650 Ma ago,[53] and an apparently unprecedentedly great deposition[54] of phosphate-rich sediments at this time.

Phosphorous provided by erosion of the Transgondwanan supermountain cannot furnish a full, straightforward explanation for all the first appearances in the fossil record listed earlier because these developments were spread out over many tens of millions of years, a period longer than that of the erosion of the supermountain. But these times actually included a double "phosphogenic" event[55] indicated by two layers of phosphorous-rich sediments with a significant hiatus between them. Thus, rather than evoking the extraordinary erosion of the supermountain as a singlet event, it is necessary to consider the several closely spaced continent-to-continent collisions that occurred during Pan-African times[56] (Chapter 1), their respective mountain-building and erosional sequels, and their culmination with erosion of the supermountain. Phosphorous may have poured into the seas and then poured into them again, before the previous load had been entirely cleared.

At about the same time, oxygen levels again apparently rose. Near the end of his *Out of Thin Air: Dinosaurs, Birds and Earth's Ancient Atmosphere*, Peter Ward found himself asking, "Could it simply be that merging continents raise oxygen levels and splitting continents draw down oxygen?"[57]

ATP

ATP is present in all living things. In early times when oxygen levels were extremely low, ATP was produced by fermentation,* still an important process in biological situations where oxygen is scarce or absent.† Fermentation produces ATP by splitting molecules of sugars, simultaneously generating carbon dioxide and a very broad variety of waste molecules. Fermentation processes are inherently inefficient because their wastes retain the potential to yield still more ATP. The type of waste produced depends on the starting material, the "food," which in today's world may be as varied as Chardonnay grapes and sewage. Waste molecules are normally excreted, but as was the case with magnetite, some may be retained within the cells that produced them.

Breathing (aerobic cellular respiration), by contrast, is 100 percent efficient, producing water and carbon dioxide and nothing else, leaving no additional ATP to be extracted.

Metazoans and the third dimension

The cells of present-day metazoans are kept together, and their tissues maintained, by numerous types of adhesive molecules, with collagen as the most important.

A 2-D biofilm is not a metazoan, nor can it ever be. The third dimension is necessary. But a 3-D mass of compatible cells stuck to one another by collagen also fails to constitute a metazoan. Animals are three-dimensional entities composed of cells that communicate and cooperate with one another. How did the third dimension come into being, and how did cells first communicate?

These days cells connect and communicate with one another through numerous types of cell-to-cell connections, structures called gap junctions among them. Gap junctions are absolutely essential for metazoan existence.[58]

Metazoans require collagen, and collagen is provided by oxygen. Metazoans also require gap junctions. But whence gap junctions?

On attempting to survive the toxic effects of oxygen, certain eukaryotic cells excreted oxygen-containing molecules. Some of these molecules, including the "oxygen expensive" molecules of the collagen family, were sticky, leaving single-celled creatures stuck to one another. This produced 2- and 3-D groups of eukaryotic cells stuck to one another, distressed and struggling to survive, but in no way cooperating.

The ocean is deep and wide, evolutionary time is long, the number of unicellular eukaryotes in each cubic centimeter of the seawater is very great, and such situations must have come into being on an astronomical number of occasions during the prehistory of the metazoa. The near-universal outcome would have been the death of the groups of cells in question, an episode in the history of life that might be preserved in the fossil record.

---

* ATP is also produced by bacterial anaerobic cellular respiration, by "breathing sulfur" and similar processes that do not concern us here.

† Yeast, an exception, prefers fermentation even in the presence of abundant oxygen, provided sugar is available.

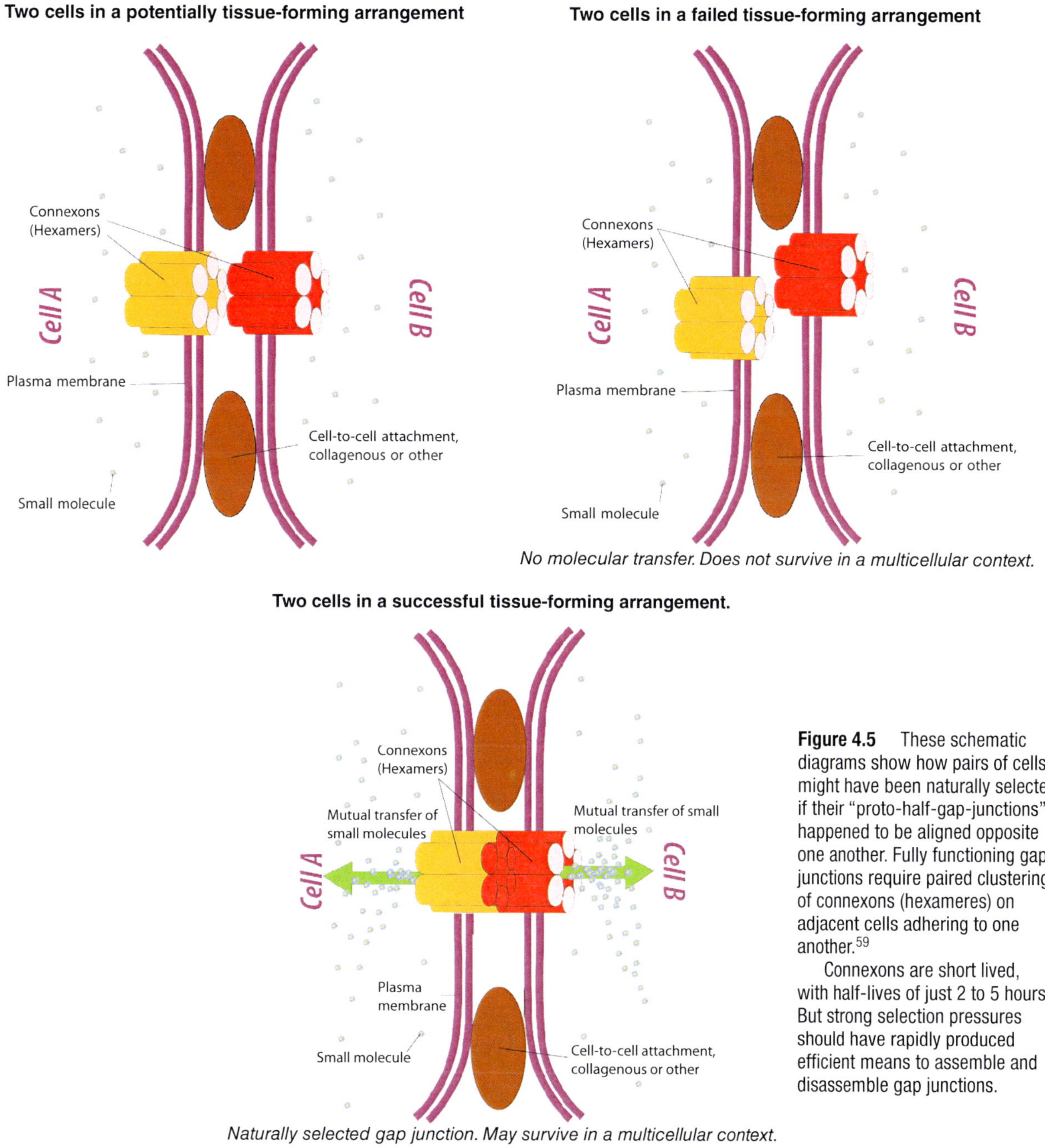

Figure 4.5 These schematic diagrams show how pairs of cells might have been naturally selected if their "proto-half-gap-junctions" happened to be aligned opposite one another. Fully functioning gap junctions require paired clusterings of connexons (hexameres) on adjacent cells adhering to one another.[59]

Connexons are short lived, with half-lives of just 2 to 5 hours. But strong selection pressures should have rapidly produced efficient means to assemble and disassemble gap junctions.

Eukaryotic cells possess channels on their surfaces through which they receive and send chemical and electrical signals to the external world. When by pure geometric happenstance certain channels in adjacent cells happened to be stuck face-to-face, channel-to-channel, they would have sent and received their signals to their neighbor rather than to the external medium. One "proper fit" may have immediately produced a gap junction between the two cells. Or there may have been a multistep period of development (**Fig. 4.5**).

We are here dealing with natural selection, the survival of those pairs of cells that were sufficiently fit to survive and reproduce. Pairs of cells with surface channels that happened to link up exchanged messages, and this exchange somehow enabled the two cells to survive, perhaps by outliving their neighbors and then consuming their remains.[60]

Collagen first, cellular interchange next (here presumed to have first been through gap junctions[61]), and then animals. A modern study recites part of my argument in its title: "Evolution of Key Cell Signaling and Adhesion Protein Families Predates Animal Origins."[62]

These days many cell-to-cell exchanges deal with the establishment of proper gradients, of getting the right amount of certain ions and small molecules into some cells and the wrong materials out.[63]

Possible sequence of events

- When exposed to toxic levels of oxygen, certain single-celled eukaryotes produced and excreted large quantities of oxygen-expensive molecules of the collagen family.
- Pairs of adjacent eukaryotes then found themselves stuck together, glued by collagen. This might have led to the emergence of the Ediacaran biota, creatures that mostly or entirely lacked full left–right symmetry (assuming the Ediacarans were indeed multicellular).
- Gap junctions came into being (**Fig. 4.5**), enabling or forcing the transfer of ions and small molecules from one cell to the next; flows and transfers could be reversed as osmotic pressure varied.
- Gradients were established (1) *within* cells and (2) *between* cells.
- Gradients *within* cells caused them to do what no amoeba or other protist ever does: to divide asymmetrically.
- With asymmetric division, one daughter cell, identical to the mother cell, provided an enduring germ line for the establishment of the species, while the other became a "tissue-former" of the individual member of the species.[64]
- Cells of the germ line produced anoxic "niches" to protect themselves from excess oxygen.[65] With time, this permitted other cells to survive and, following generations of natural selection, there emerged the specialized tissues of muscle, liver, or bone, each composed of cells tightly constrained in its ability to proliferate or vary.
- Gap junctions also regulate the oxygen gradient *between* cells within tissues[66] so that deadly levels cannot accumulate. ("The dose makes the poison.") Thus, functioning oxygen gradients must have been established from the outset.
- Each phylum came into being with a genetic program that stabilized the particular gradients for the construction of their characteristic body plan. These gradients set initial conditions that imposed lasting geometric constraints on the body-plans of the animals of each phylum.[67]

Collagen and cancer

Cancer is a pathology of three-dimensional multicellular animals. In individual instances, its cause, if determinable, may be traced to heredity, viral infection, damage by ionizing radiation,[68] physical damage by fibers of asbestos, chemical damage, damage caused by smoking, chronic inflammation (as opposed to acute, non-chronic inflammation), disease, injury, age, inadequate immunity, loss of access to adequate supplies of oxygen, or other circumstances. The factor common to all such triggers is the ability to inflict mechanical damage to a tissue, *micromechanical* damage that produces *architectural tissue-defects* and causes the tissue to lose full or partial control of one or more of its component cells. Cancer can be understood as due to the failure of collagen to maintain the integrity of a tissue, thereby hindering gap junctional (and other) cell-to-cell communications that regulate growth, the differentiation of cells into the specific types suitable for particular tissues, and the triggering of cell death (apoptosis) when required.

Cancers may, for example, be caused by the insertion of asbestos fibers of particular sizes between individual cells. This may lead to loss of cell-to-cell communication, leaving some cells starved for electrical and chemical signals and for oxygen, and partly or entirely isolating them from their home-tissue. Fibers may also change the plane along which cells divide, which may produce cumulative effects with each subsequent cell division and thereby geometrically distort the tissue.[69]

If not destroyed by the immune system, or dying "of natural causes," uncontrolled cells, tightly tailored ("differentiated") for particular tasks within their tissues, would struggle to survive in their altered surroundings. They would do whatever they could in the absence of some of the many supports normally supplied (and constraints imposed) by their tissues.[70]

One strategy for cells deprived of access to the exquisitely dosed supply of oxygen required by tissues to maintain moment-to-moment ATP requirements (but no more) would be to revert to their archaic ability, common to many animals,[71] to maintain themselves by fermentation. This is a critical step toward cancer for, as understood by the Nobel Prize winner Otto Warburg (1883–1970), "cancer has countless secondary causes" but

> ... only one prime cause. Summarized in a few words, the prime cause of cancer is the replacement of the respiration of oxygen in normal body cells by a fermentation of sugar.
>
> — Otto H. Warburg[72]

Warburg also discovered that cells that revert to anaerobic respiration cannot later be restored to health by reinstating the supply of oxygen. Reinstatement of the supply of oxygen does not reverse the course of cancer, perhaps because instructions to cancer cells to self-destruct can no longer be sent or received. But, whether this suggestion has merit or not, the fact is that cancer cells that had been partly released from their tissue-constraints and then restored to their tissues, would no longer be the same. They would be larger because of their retention of fermentation products.[73] This would produce 3-D geometric distortions in their tissues. Zones of weakness would develop, especially at bends and at the joins between tissue types or organs. Tissues might then lose control of yet other cells, and so on.*

The original function of the earliest immune systems may have been to attack cells that in one way or another refused to respect tissue constraints, perhaps as signaled by certain products of fermentation. The logic here is not that of a biologist. It is the logic of history for, "historically speaking," the very early metazoan tissues are likely to have been threatened by recalcitrant cells before ever being threatened by infectious agents.

The original function of inflammation may have been mechanical, to keep injured tissues from coming apart. Yet inflammation, as indicated by the word itself, is also characterized by heat, as, too, is fever. Neither heat nor fever has a mechanical role but cancer cells are significantly more sensitive to heat than are healthy cells. This may be why the very rare individuals who undergo seemingly spontaneous remissions or cures of their cancers often report having undergone "a hefty feverish infection" somewhat before.[74] Immune system, inflammation, and fever[75] may have all originated as mechanisms to maintain the integrity of living tissues.

Cells released from the constraints imposed by their tissues may survive and proliferate. In addition, they will vary. Proliferation and variation. This is just a rewording of the Darwinian "descent with modification." In the context of cancer, variation may lead to the production of some very peculiar cells indeed — ask any pathologist — and many such cells have characteristics that are incompatible with the metazoan requirement that individual cells cooperate with one another.[76]

Once a solid cancer is established, a continued or acute inadequate supply of oxygen rapidly reduces the adhesion of individual cancer cells, without, however, killing them.[77] Liberated cancer cells may then travel within the individual metazoan. Anchored at other tissue-sites by unknown mechanisms, perhaps facilitated by their extreme variation, descendants of these cancer cells may cause secondary cancers (metastases).

---

* Direct observation of cancerous tissues under the microscope shows some cancer cells that are smaller, rather than larger, than healthy cells. Diminished cell size would also cause 3-D tissue-distortions but the reason for the reduction in size, if confirmed, is not known, though cell starvation may be a possibility.

Calcium and calcium detoxification

By sometime before 542 Ma, adequate supplies of phosphorus, hence of ATP, had permitted animals to grow larger. But during the interval from 543 to 515 million years ago, the calcium content in seawater increased by a factor of three,[78] and calcium is a toxic substance. Indeed, "the calcium ion, Ca^{2+}, is pharmacologically one of the most disruptive substances for normal cell function," and intracellular concentrations of calcium ions are carefully regulated.[79] Saltwater biofilms of the cholera bacterium, *Vibrio cholerae,* disintegrate when a calcium-binding compound is added to their environment, an effect that does not occur when ions other than those of calcium are similarly bound.[80] Expressed more broadly, "calcium makes germs cluster,"[81] the same defensive mode adopted by anaerobes exposed to oxygen (**Fig. 4.4**).

A "toxic ion that must be removed from most cells,"[82] calcium accumulates extracellularly,* "and the occurrence of calcium deposits may therefore represent a form of detoxification."[83] Hence "biomineralization may be a cellular detoxification mechanism."[84]

When oxygen and calcium are simultaneously present in toxic amounts, the two may be inexpensively excreted together as, for example, in seashells where the prime constituent is calcium carbonate, $CaCO_3$ (the minerals aragonite and calcite).

A survey of calcium deposition in diverse tissues from various marine invertebrates also detected aluminum, cadmium, cobalt, lead, silver, and 10 other potentially toxic substances within calcium-rich granules.[85] These substances, thus removed from metabolic functions, were present in widely different ratios from one granule sample to another, apparently reflecting the toxins to which individual creatures had been exposed.[86]

The evident adaptability of the ancient detoxification mechanism suggests that it may have been co-opted for additional biological purposes as well and likewise, that its dysfunction may be implicated in pathologies other than cancer. It is also consistent with the suggestion that the original role of hemoglobin had been to scavenge oxygen.[87]

Similar reasoning indicates that the bones of vertebrates, which are primarily composed of calcium phosphate,† also had their ultimate origins as the products of detoxification in times when phosphorus was readily available.

Bio-parts of silica exist but are rare. Some Demospongiae, a class of sponges that have no common name, have skeletons composed of the collagen-like protein spongin with spicules made of silica; the spongin, and apparently the silica too, are excreted and the spicules grow extracellularly.[88] Based on observations of these and other sponges, specialists have ventured that "there may be some common mechanisms" in the rather different systems of mineralization for silica and calcium.[89]

In recent years, better-understood features of these mechanisms have been harnessed in the manufacture of hybrid materials for biocompatible medical implants, with collagen used as a template.[90] In nature, in the lab, and in industry, "collagen" is well suited for use as a template or guide for the deposition of other materials, a consequence of the repeating structure of molecules of the collagen family. These serve as regularly spaced anchor-points, different types of collagen providing different spacings.

Conservation of biomineralizing pathways

When marine animals first began to manufacture shells of calcium carbonate, some precipitated aragonite and others precipitated calcite, two minerals with the same chemical composition, $CaCO_3$, but different crystal structures. Animals would precipitate whichever then required less energy, a matter determined by the composition and temperature of the seawater at the time.[91] Yet despite subsequent changes to the chemistry of the seas, groups of animals "rarely switched mineralogies,"[92] thus indicating the conservative nature

* As does collagen and, judging from experience at the Siljan drill-hole, magnetite (*page 111*, sidebar).

† $Ca_5(PO_4)_3(OH)$, the mineral hydroxylapatite.

of mineralizing pathways as regards calcium. Comparable pathway conservation may have been maintained for phylum-specific varieties of collagen, and other molecules. In investigating such matters, it will be useful, and perhaps necessary, to relate the secretion of each substance to the "myriad biochemical mechanisms" employed by organisms to detoxify reactive oxygen species and derivatives.[93]

Final remarks

Many of the events involved in the emergence of complex life have been driven by the need of individual cells to rid themselves of toxic excesses of oxygen and calcium. This involved the concurrent shedding of other elements that happened to be abundant where and when oxygen or calcium reached toxic concentrations, iron, silicon, phosphorus and carbon, in particular. The shedding into extracellular space of oxygen-expensive molecules of the collagen family induced 3-D tissuelike structures to come into being, necessitated cell-to-cell cooperation, and led to the emergence of the metazoa.

I had a discussion with Théodore Monod (1902–2000) toward the end of his long life. We were deep in Egypt's Western Desert with time to talk while our driver-mechanic was making a repair. "Professor Monod," as I always addressed him despite four decades of friendship, was author or coauthor, or the subject, of an astonishing 2167 publications, many of them descriptive studies of highly specific scientific topics — "On a new species of cavern dwelling shrimp," for example (Monod, bibliography, entry #1385), or "Morphological anomaly in *Acacia tortilis,* of Mali" (#1746).[94]

Monod had been among those whose self-assigned task, the task recommenced during the Renaissance, was to describe the world. But that day in the Sahara, I tried to convince him that for the next generation of scientists, the job would be to construct a grand synthesis. I failed. He was not convinced. There were still, "thank goodness," he said, endless things about which we knew nothing. The vehicle was then ready and we never returned to our conversation.

Notes and references – *Chapter 4: Animals and Cancer*

1 Chatterjee et al. (2013); Feneyrol (2012).
2 Sections in the remainder of this chapter are revised versions of my papers "Origin of the Phyla and Cancer" (2007b), and "Did Detoxification Processes Cause Complex Life to Emerge?" (2009).
3 Des Marais et al. (1992).
4 Nursall (1959).
5 Huldtgren et al. (2011); Lipps, Stoeck, and Dunthorn (2012).
6 Chen et al. (2004).
7 Elsewhere I argue that these may have grown directly from blastulas without undergoing full or familiar gastrulation-style cell rearrangement (Saul, 2007b).
8 Clites, Droser, and Gehling (2012).
9 Seilacher et al. (2003); Seilacher (2007).
10 Rogov et al. (2012).
11 Matz et al. (2008).
12 Dzik (2005).
13 Dzik (2005).
14 Dzik (2005).
15 See Parcha et al. (2005).
16 Dzik (2005).
17 These tube fossils, *Onuphionella* and *Spiroscolex*, were apparently lined with a somewhat sticky material to which mica flakes, but not heavier mineral grains, could adhere.
18 Knoll and Lipps (1993).
19 Ridgwell et al. (2003).
20 Porter (2007).
21 Fossil bryozoa ("moss animals") have not been found in rocks older than 485 to 480 Ma, but the bryozoa constitute a bona fide phylum. When phylogenetic analyses are performed on living bryozoans, they "never fall out within another recognized group/phylum" (Kenneth M. Halanych, personal communication, 8 November 2009).
22 Signor and Lipps (1992).
23 Simkiss and Wilbur (1989).
24 Signor and Lipps (1992); Culver (1991); Signor and Ryan (1993).
25 Kirschvink and Hagadorn (2000) p. 143.
26 Kirschvink and Hagadorn (2000).
27 John Harshbarger, personal communication (30 October 2002). Benign and malignant tumors have been recognized in dinosaur bones (*Science News*, 28 August 1999, p. 141).
28 For the prevalence of cancer in humans and recently domesticated creatures such as hybrid ducks, laboratory mice, domesticated trout, Lipizzaner horses, cats and dogs, especially large dogs such as great danes and boxers, and for its infrequent occurrence among wild animals and among members of less complex phyla, see Graham (1992) and Saul and Schwartz (2007). I thank Annerose Anders for useful personal communications in 2002 on this matter.
29 The Platyhelminthes (flat worms) get their oxygen and nourishment by diffusion through their bodies, and their body plans may approach the greatest complexity allowable for animals lacking body cavities. For cancerous lesions in Platyhelminthes, see Harshbarger and Gibson (1982). John Harshbarger (personal communication, 29 October 2002) wrote: "Chordate cancer is abundant in the top 5 classes and present in the lower 3. Among invertebrates it is only abundant in one class of Mollusca (Bivalvia)." For reasons why more complex and recently evolved species and breeds are particularly susceptible to cancer, see Graham (1992).
30 Saul (1994, 2007b, 2008); Saul and Schwartz (2007).
31 Kump and Barley (2007).
32 Kump and Barley (2007).
33 Fenchel and Finlay (1995).
34 Gold [ed. by Mitton] (2012).

35 Some eukaryotic algae (euglenoid flagellates) contain several thousand submicron sized magnetite crystals; Kirschvink and Hagadorn (2000) p. 143. Arctic foxes should probably also be added to the list of animals possessing an ability to sense the Earth's magnetic field http://www.edisproduction.de/2013/11/19/fox-hunting-under-snow-in-an-incredible-way/ (Retrieved 22 December 2013).
36 Vali and Kirschvink (1990) p. 97, as cited by Kirschvink and Hagadorn (2000) p. 145.
37 Cited by Harris (2013) in his "Evolution's Other Narrative: Why Science Would Benefit from a Symbiosis-Driven History of Speciation."
38 Towe (1970) p. 781.
39 Towe (1981).
40 Towe (1981).
41 Towe (1981); Saul and Schwartz (2007).
42 Des Marais et al. (1992); Squire et al. (2006).
43 Knoll et al. (2006).
44 Canfield et al. (2007). The end of the Gaskiers glaciation, 582–580 Ma, probably produced an overturning circulation of the oceans with consequent oxygenation of the deep ocean.
45 Saul and Schwartz (2007).
46 Saul (2007b).
47 This suggests that cells of the Ediacara, if they were indeed multicellular, did not possess fully functioning HOX genes.
48 See Squire et al. (2006).
49 Squire et al. (2006).
50 Squire et al. (2006); also see Sankaran (2007).
51 Squire et al. (2006) p. 127.
52 Squire et al. (2006).
53 Squire et al. (2006).
54 Squire et al. (2006). Referring to the simultaneous flux of strontium into the oceans, Derry (2006), as cited earlier, asked "If continental collisional events alone are responsible for imparting high $^{87}Sr/^{86}Sr$ ratios to the oceans, why haven't we seen episodic fluctuations to high values before 600 million years ago, associated with known earlier collisions?"
55 Brasier (1990).
56 Veevers (2003).
57 Ward (2006) p. 233. Compare with Touret and Huizenga (2011), cited at the end of Chapter 1.
58 Revel (1988), whose work was not seen, was the first to understand this. Trosko (2011); http://php.med.unsw.edu.au/cellbiology/index.php?title=Cell_Junctions
59 Connexin proteins are coded by twenty or more highly evolutionarily conserved connexin genes (Trosko (2011) p. 58).
60 A similar mechanism may be at work among the cells in very early embryos (Saul, 2007b).
61 Gap junctions are unique among the many types of cell-to-cell junctional complexes in their lack of strong or systematic linkages to actin fibers or other cytoplasmic elements.
62 King et al. (2003).
63 Sheridan (1987).
64 Trosko (2007, 2011); Kang and Trosko (2011).
65 The origin and essence of this niche was geometrical; see Saul (2007b).
66 Lowenstein (1979).
67 Gap junctions are also "involved in the early generation of left–right asymmetry," and in permitting the construction of animal bodies with asymmetrical internal organs (Levin and Mercola, 1998).
68 Unicellular creatures are subject to viral infections and damage by ionizing radiation, but the resultant pathologies are not cancers by any familiar definition.
69 Schwartz et al. (2002).
70 Sonnenschein and Soto (1999); Saul and Schwartz (2007).
71 Budd (2008).
72 Warburg, "The Prime Cause and Prevention of Cancer," http://healingtools.tripod.com/primecause1.html/ (accessed 7 March 2013).

73 Schwartz (2004). Noting that land vertebrates spend one-third of their lives sleeping, Graham (1992) pp. 102–104, 143, asked "what killed animals that did *not* sleep?" and concluded that "sleep's primary function is to defend against cancer." His assertion was reinforced by recent research that demonstrated that waste molecules produced in the brain during waking hours are flushed out by greatly increased flow of cerebrospinal fluid during sleep; see http://www.nytimes.com/2014/01/12/opinion/sunday/goodnight-sleep-clean.html?emc=eta1&_r=0

74 Hobohm (2005). This thought-provoking review, "Fever Therapy Revisited," suggests that promising paths of cancer research are not at present receiving sufficient attention.

- *Item:* "Cancer patients usually carry… T-cells that recognize tumor cell antigens," and it is not "only that tumors are invaded by defense cells but also that the number of invading cells and survival rate can correlate…." (citing research results dating to 1958).
- *Item:* Hobohm assumes that cancer regression is the result of "full-blown immune attack." In making this assumption he refers to observations and research from 1868, 1891 and 1918 that seem to be more pertinent than much of the recent research he surveys.
- *Item:* Many studies show a correlation between cancer remission or cure and "high temperatures of several days duration."
- *Item:* "Fever, in particular, during the second half of the 20th century, which is characterized by heavy use of antibiotics, was regarded unequivocally by main-stream medicine as an unnecessary, weakening state which should be avoided or prevented."
- *Item:* There is "a striking inverse correlation between the number of infections and mortality from tumors in Italy in the period 1890–1960: every 2 percent reduction in the number of infectious diseases was followed by a 2 percent increase in tumors about ten years later." But since "fever is metabolically expensive, it must provide substantial advantage to the host. … Post-operative infections can prolong survival…. In this light, it seems unfortunate that fever is usually suppressed in hospital routine." Hobohn provides further information and detailed references at http://www.nature.com/bjc/journal/v92/n3/full/6602386a.html.

It is unclear whether activation of the immune system or heightened body temperature by itself is the more important factor in combating cancer.

75 Normal responses to heat shock enable healthy cells to withstand elevated temperatures for limited periods. Lack of oxygen may suppress such responses.

76 Saul (1994); Sonnenschein and Soto (1999); Saul and Schwartz (2007).

77 Hasan et al. (1997).

78 Brennan et al. (2004).

79 Simkiss (1977) p. 199.

80 Kierek and Watnick (2003).

81 Harder (2003) p. 293.

82 Simkiss (1977) p. 199.

83 Simkiss (1977) p. 199.

84 Simkiss (1977) p. 199.

85 Cited by Simkiss (1977).

86 Simkiss (1977), p. 199, suggests that for many metazoans, sequestration of calcium by precipitation in the form of highly insoluble intracellular granules "may be energetically more economical than pumping it out of the cells into a supersaturated body fluid."

87 Minning et al. (1999).

88 Müller et al. (2006).

89 Simkiss and Wilbur (1989) p. 143.

90 Heinemann et al. (2007) p. 1.

91 Porter (2007). Hazen et al. (2013), p. 35, write that "Among the most fundamental mineralogical questions — ... evident from any museum display of carbonate minerals — is what physical, chemical, and biological processes lead to the remarkable range of calcite crystal forms. No other crystalline phase exhibits such a wide range of morphologies. What environmental factors influence calcite crystal forms? And why don't other rhombohedral carbonates display similar variety?"

92 Porter (2007) p. 1302.

93 Raymond and Segrè (2006) p. 1764.

94 Titles translated from French; see Hureau and Escudier (2005).

5

The Ends of Darwinism

DARWINIAN PRINCIPLES apply to all forms of life.

There are three extreme cases, however, for which reasoned claims can be made that the principles set out by Darwin do not fully apply. The first, which is hypothetical, is discussed in Tom Gold's *Deep Hot Biosphere*. Gold argued that life started at depth using hydrocarbons, not the Sun, as a source of energy. In this view, life migrated to the Earth's surface at a much later time.[1]

The deep domain in which life arose, according to Gold's supposition, is exceedingly stable and in places virtually unchanging. With no environmental change, there would be no pressure to adapt, and no natural selection. In consequence, some of the very early forms of unicellular life might be exceedingly stable, living at depth in astronomical numbers within rocks worldwide, perhaps dormant or encysted but unchanging and not dying, reproducing (dividing) on a time scale of perhaps billions of years, and seemingly immortal. Subduction and other geological changes might be as inconsequential as passing storms for populations of species such as these.

Ignoring the many assumptions and uncertainties involved in such follow-ups to Gold's thesis, a hypothetical population of identical 4 billion-year-old unicellular creatures would still not be exempt from the workings of natural selection. Such creatures, if they exist, would inhabit an extraordinary niche in which no significant environmental change has *yet* occurred. When it eventually does, next week or in a few billion years, members of these hypothetical populations might adapt to the new circumstances.

Adaptation requires variation, which would necessarily exist no matter how hypothetically stable the environment and the genome, an inescapable real-world consequence of occasional radiation damage. In short, there has never been, and never can be, a truly unchanging niche or an absolutely stable genome, a "fact of life" observed during an ongoing 25-year experiment in bacterial evolution.[2]

Two things make such populations special, or would make them special if they exist. One is the exceptional time scale. The other is more subtle, for whereas Darwinism involves "descent with modification," radiation is able to produce variation, that is, modification, within a single generation.[3]

Charles Darwin had no explanation for the sudden appearance during Cambrian times of the varied multitude of multicelled animals. Anti-Darwinians subsequently fixed on the seeming inability of natural selection to explain the Cambrian Explosion as though it were a fundamental flaw in the Darwinian edifice, and as though it automatically called for something akin to divine intervention.

In the previous chapter, I argued that when oxygen rose to toxic levels during the run-up to the Explosion, certain single-celled creatures excreted oxygen-rich molecules, collagen in particular, that caused them to stick to one another, and that natural selection then rapidly, even "explosively," led to the emergence of multicelled animals.

If I am correct, I have undone a key argument of the Creationists, of those who imagine a need for an Intelligent Designer to explain the existence and the varied body plans of the world's first multicelled creatures. But *even if I am wrong* on the matter of oxygen and collagen, I will have accomplished much the same thing. For I have shown that a straightforward instance of natural selection, of Darwinian reasoning, whether right or wrong in its details, is capable of explaining the sudden appearance of multicelled animals without recourse to the will, existence, or acts of an Intelligent Designer or Creator.

The third special case concerns humans.* Humans are different from other animals and it does not require a religious mindset to see this truth. Our level of existence has a qualitatively greater complexity to it.

In his "Evolution and Tinkering," which has my vote as the best scientific paper ever, François Jacob (1920–2013) set out the notion of the hierarchy of objects and systems. His main example was physics, chemistry, biology, sociology. "At each level, new properties may appear which impose new constraints on the system," Jacob explained.

> But these are merely additional constraints. Those that operate at any given level are still valid at all more complex levels.... [E]very proposition that is valid for biology holds true in sociology. But as a general rule, the statements of greatest importance at one level are of no interest at the more complex ones. The law of perfect gasses is no less true for the objects of biology or sociology than for those of physics. It is simply irrelevant in the context of the problems with which biologists, and even sociologists, are concerned.[4]

Darwinian principles apply to us, fully so,[5] as does the law of perfect gases, but it is not these matters that make us different or interesting. It is our complex non-animalian ways that make us interesting.

What is necessary, Jacob understood, is to work out the rules for each particular level. Jacob gave a hint how to start: "Simpler objects [those of physics in his example] are more dependent on constraints than on history. As complexity increases [as is the case of biology], history plays a greater role."

The story of civilization, of the ways of humans, of society, is more complex than biology, and history is even more at play. So let us consider the very first historical event, the very first thing we did that had a complexity of a sort never employed by other creatures: *we shared complex ideas. It follows that the development of civilization, which is to say history, is all that which has been impelled by such shared thoughts.*

Questions arise. Why did we do this? And what were the first complex ideas to be shared, the initial condition or conditions, of our humanness?

As I have argued in *The Tale Told in All Lands*, the first idea to be shared was a question that all have posed: What shall we do about death? From a Darwinian point of view, posing the question and discussing possible ways by which death might be delayed or avoided is just another survival strategy. For those who adopt Jacob's approach, however, it is the essential step up to the next level, the emergence of conscious cooperation in the struggle to survive.

In time a strategy was developed. It can be called religion, or it can be called science. Humans would identify undying aspects of their surroundings. They would then try to determine what caused these particular things not to die or to be reborn. And once the

---

* Most of the remainder of this chapter has previously appeared in *The Tale Told in All Lands*, Les 3 Colonnes, Paris, 2013. http://www.tomebook2.com/

cause or causes were identified, great efforts would be made to incorporate them into the ways of humans.

In times past, and even now, the availability of many types of nourishment was linked to the renewals of the seasons and the return of vegetation, fish, and game, and the birthing of certain animals. These were life-giving effects. But where were their causes?

Our earliest written records come from the Sumerians of ancient Mesopotamia, whose writings reflect the beliefs of still earlier peoples unknown to us.[6] For Sumerians, the "informing thought" of the "world feeling" was "What is above is below."[7] Similarly, the Chinese held that "everything terrestrial" had "its prototype, its primordial cause, its ruling agency in heaven."[8] In Europe the same idea prevailed in the most familiar saying of the Western alchemists: "That which is above is like that which is below and that which is below is like that which is above."[9] And at least one influential European sect founder would proclaim as the "Great Axiom" that "The Above" was "the exact mirror image of the Below."[10] Although these widely separated traditions are usually regarded as distinct entities, they in fact share a common element. Poorly understood backwaters of the Western heritage and a major tributary of Chinese culture tap the same source as does the mainstream of Sumerian thought. The concept appears in other cultures as well. In the Jewish *Zohar*, which dates to the 13th century but draws on much older sources, we find that "the inferior world is a reflection of the superior." Similarly, the Micmac Indians of the Canadian Maritime Provinces hold that "In all things as it was and is in the sky, so it is on earth,"[11] while in Colombia, many Indian cultures "conceptualize the sky as a blueprint for past, present and future occurrences on earth..."[12]

Effects in the world Below were identified with *causes* Above. Hence, astronomy became the first science, and the heavens Above became the domain of the immortal and supposedly immortalizing gods.

Thus it is that Mercury, Venus, Mars, Jupiter, and Saturn come down to us as the names of gods, of immortal gods. And thus it is that Mesopotamian data concerning certain of their periodicities (their "behaviors," their "renewals") were virtually as good as our own before the advent of high technology. The secret of the ancient gods was not omnipotence. Zeus-who-is-Jupiter could not even manage his domestic affairs. The secret of the gods was immortality.

The great innovation of the Mesopotamians was Sacred Kingship, which, as we are informed by cuneiform texts, had been "lowered from heaven." With Sacred Kingship, church and state, science and religion, were one.[13] The King would receive the life-giving essence of heaven, or so it was held, and spread it upon the people. With Sacred Kingship, the foundations of all later complex societies were laid.

The King and his Sacred Court (and their predecessors before the advent of Kingship) sought a mechanism that would allow earthbound mortals to obtain the essence of the supposedly ever-constant, ever-returning heavens. But in time, it was understood that the Heavens are not constant after all. They gradually change as a consequence of the Precession of the Equinoxes. Not everyone knew, but the court astrologers did and, terrified, found that their livelihoods and lives depended on devising ways to update the manner by which the heavens were to be projected earthward. A new Age would be announced, a new beginning. Or it would be discovered that the Mandate of Heaven had changed. And if the pole star (whose position slowly changes with the Precession) was taken as the Center of Heaven, as it was by many peoples, great efforts might be made to discover the new "Center of the Earth" where the King should reside. When all else failed, a great resetting might be retroactively installed as it was after the Trojan War (of myth) and the (mythic) founding of (the real) Rome, or by elusive "Inceptions" such as those that mark the beginnings of some Chinese dynasties. Calendrics and correlations, the founding of new cities and the establishment of new borders, rituals and rules, bloodlines and birthstones, were all involved in renewed attempts to render the Earth "as Above."

The original project was not abandoned. Initial conditions are rarely cast away. Instead it was endlessly modified: cities represented stars, myths and poetry told the skies, mathematics tracked celestial goings-on, and laws and wars enabled the King so that he might receive the beneficence of heaven. *The intended goal was immortality. The unintended result was civilization.*

In the beginning, all had been for the best, for man's invention of civilization, founded in logic and demanding observation of the world around him had enabled people to live longer, longer and better. But the wonderful accidental invention, civilization, would be degraded by the very enemy it had been designed to vanquish, by Time itself, as violent nationalisms and intolerant religions emerged from ill-adapted institutions designed by the well-intentioned to attain the unachievable.

Direct joustings with the windmill of death belong to the past. The time has come to comply with Teilhard de Chardin's admonishment to "lay aside the ancient prejudices and build the earth." These days researchers investigate the aging process at the molecular level. It thus seems likely that a sustained following of Teilhard de Chardin's advice, in biology at least, might produce ways of slowing the ravaging effects of time. We have good reasons to hope that knowledge may permit us to regulate Time's Mill so that it will grind less rapidly.

But in many fields, modern researchers have made little progress in undoing the insubstantive plausibilities and artificial certainties that are our heritage. In consequence, as put by H.G. Wells, a race is on between education and catastrophe. Painful education or still more painful catastrophe, we do not know which will prevail. The archaic elements within our heritage may bring catastrophe to us all, whether through un-modern aversion to certain types of knowledge, a predilection to obey those whose epaulettes carry a star or the eagle that once belonged to Zeus, or through lingering faith that a solution to the human predicament may yet be extracted from an uncomprehended and incomprehensible formulation of the ancient kind. Those who still hope that their destiny can be manipulated by fashioning heaven on earth may ponder the mocking questions addressed to Job:

> Tell me, since you are so well informed!...
> Have you grasped the celestial laws?
> Could you make their writ run on the earth?
> — Job 38:4, 38:33 (*Jerusalem Bible*)

It was not always so, but "As Above, So Below" is nowadays and has long been dangerous nonsense. T.S. Eliot, who was a Believer throughout his life, formulated the enduring puzzlement in its exact historical terms. Addressing both Believers and nonbelievers, he agonized whether our forebears had deceived us. Or whether, "deceived themselves," they had unknowingly left us "merely a receipt for deceit."

Notes and references – *Chapter 5: The Ends of Darwinism*

1 Gold (1999).

2 See Pennisi (2013b). Depending on the boundaries of this hypothetically near-perfect niche, horizontal gene flow from neighboring unicellular creatures would be another potentially confounding factor.

3 "DNA sequences of hydrogen-eating microbes extracted from rock fractures deep below North America, Europe, South Africa and Japan" have been found to be approximately "97 percent similar – making them virtually the same species." http://www.independent.co.uk/news/science/life-on-earth-may-have-developed-below-rather-than-above-ground-reveal-scientists-8991601.html (16 December 2013). Here I emphasize the 3 percent difference.

4 Jacob (1977) p. 1162.

5 In one modern context, the fittest are those who have access to the best medical care.

6 T.B. Jones (1969) p. 139, n.1, citing and paraphrasing A. Leo Oppenheim; also see E.A. Speiser in Jones (1969) p. 102.

7 Alfred Jeremias, as cited by J. Campbell (1974) p. 87.

8 Eitel (1979) p. 10. The Taoist text, the *I-Ching*, notes that the sage "...looking up, contemplates the brilliant phenomena of the heavens and looking down, examines the definite arrangements of the earth.... He traces things to their beginning and follows them to their end.... [T]hus he knows what can be said about death and life...."; see Bulling (1952) p. 12.

9 Alchemical citation from *The Emerald Table*; see Yates (1978) p. 150.

10 Lanz von Liebenfels (1931) p. 106. The great historical influence of the little-known Lanz is discussed by Daim (1958), and by Saul (2013).

11 Hagar (1900) p. 95.

12 Gerardo Reichel-Dolmatoff, as cited in Opperman (1981) p. 5.

13 "Sacred Kingship" in the *Encyclopaedia Britannica* (15th ed.) provides a good introduction to the subject.

Bibliography

Austromineral Ges.m.b.H., Vienna (Oct. 1978): *Geological Prospecting and Economic Assessment of the Gemstone Belt in Southeastern Kenya*, Kenya-Austria Mineral Exploration Project, Second Report to the Ministry of Natural Resources, Mines and Geological Department, Nairobi.

Balmer, Walter Adrian (2011): "Petrology, Geochemistry, and Gemmological Characteristics of the Marble-Hosted Deposits of the Morogoro Region, Tanzania," Ph.D. thesis, Chulalongkorn University, Bangkok, 184 pp.

Bear, L.M. (1952): "A geological reconnaissance of the area south-east of Embu," *Report, Geological Survey of Kenya*, no. 23.

Beloussov, V.V. (24 Apr. 1979): "Why do I not accept plate tectonics?" *EOS Trans. American Geophysical Union*, vol. 60, no. 17.

Bendick, R., and R. Bilham (2001): "How perfect is the Himalayan arc?" *Geology*, vol. 29, no. 9, pp. 791–794.

Bertrand, Guillaume, and Claude Rangin (2003): "Tectonics of the western margin of the Shan plateau (central Myanmar): Implication for the India-Indochina oblique convergence since the Oligocene," *Jour. Asian Earth Sciences*, vol. 21, pp. 1139–1157.

Bjørlykke, Knut (1981): "Late Precambrian tillites of the Bunyoro series, western Uganda," In: *Earth's Pre-Pleistocene Glacial Records*, W.B. Harland and M.J. Hambrey, eds., Cambridge University Press, pp. 151–152.

Bleeck, A.W.G. (1908): "Rubies in the Kachin Hills, Upper Burma," *Records, Geological Survey of India*, vol. 36, part. 3, pp. 164–170.

Boswell, Ray, and Timothy S. Collett (2011): "Current perspectives on gas hydrate resources," *Energy & Environmental Science*, vol. 4, pp. 1206–1215. DOI: 10.1039/c0ee00203h.

Bottke, William F., David Vokrouhlický, David Minton, David Nesvorný, Alessandro Morbidelli, Ramon Brasser, Bruce Simonson, and Harold F. Levison (3 May 2012): "An Archaean heavy bombardment from a destabilized extension of the asteroid belt," *Nature*, vol. 485, pp. 78–81. DOI: 10.1038/nature10967.

Bowerbox, Gary W., and Bonita E. Chamberlin (1995): *Gemstones of Afghanistan*, Geoscience Press, Box 42948, Tucson, Ariz.

Boyle, Robert (1672): *An Essay About The Origine & Virtues Of Gems. Wherein are Propos'd and Historically Illustrated some Conjectures about the Consistence of the Matter of Precious Stones, and the Subjects wherein their chiefest Virtues reside,* William Godbid, London.

Brain, C.K. ('Bob'), Anthony R. Prave, Karl-Heinz Hoffmann, Anthony E. Fallick, Andre Botha, Donald A. Herd, Craig Sturrock, Iain Young, Daniel J. Condon, and Stuart G. Allison (2012): "The first animals: ca. 760-million-year-old sponge-like fossils from Namibia," *South African Jour. Science*, vol. 108, no. 1/2, Art. #658, 8 pp. DOI: org/10.4102/sajs.v108i1/2.658.

Brasier, M.D. (1990): "Phosphogenic events and skeletal preservation across the Precambrian-Cambrian boundary interval," In: A.J.G. Notholt and I. Jarvis, eds., *Phosphorite Research and Development, Special Publication, Geological Society,* London, vol. 52, pp. 289–303.

Brennan, Sean T., Tim K. Lowenstein, and Juske Horita (2004): "Seawater chemistry and the advent of biocalcification," *Geology*, vol. 32, no. 6, pp. 473–476. DOI: 10.1130/G20251.1.

Bridges, Campbell (2007): "Le grenat vert tsavorite: presentation et débat," *Révue de Gemmologie, a.f.g.,* no. 161, pp. 4–7.

Brothers, L.L., C.L. Van Dover, C.R. German, C.L. Kaiser, D.R. Yoerger, C.D. Ruppel, E. Lobecker, A.D. Skarke, and J.K.S. Wagner (2013): "Evidence for extensive methane venting on the southeastern U.S. Atlantic margin," *Geology*, vol. 41, pp. 807–810.

Brown, Michael (2008): "Characteristic thermal regimes of plate tectonics and their metamorphic imprint throughout Earth history: When did the Earth first adopt a plate tectonics mode of behavior?" *Geological Society of America, Special Paper* 440.

Budd, G.E. (2008): "The earliest fossil record of the animals and its significance," *Phil. Trans. of the Royal Society (B) Biological Sciences,* vol. 363, pp. 1425–1434.

Bulling, A. (1952): *The Meaning of China's Most Ancient Art*, E.J. Brill, Leiden.

Bunge, H.-P., and S.P. Grand (2000): "Mesozoic plate-motion history below the northeast Pacific Ocean from seismic images of the subducted Farallon slab," *Nature*, vol. 405, pp. 337–340.

Byler, William H. (1983): "Circular Structures of Earth," In: American Society of Photogrammetry, 49th Annual Meeting (March 13–18, 1983, Washington, D.C.), Technical Papers, ASP-ACSM (A84-33326 15-43), pp. 471-480, Falls Church, Va.

Camacho, A., B.J. Hensen, and R. Armstrong (2002): "Isotopic test of a thermally-driven intraplate orogenic model," *Geology*, vol. 30, pp. 887–890.

Cameron, A.G.W., and W. Benz (1991): "The origin of the moon and the single impact hypothesis IV," *Icarus*, vol. 92, pp. 204–216.

Campbell, Ian H., and Richard J. Squire (2010): "The mountains that triggered the Late Neoproterozoic increase in oxygen: The second great oxidation event," *Geochimica et Cosmochimica Acta,* vol. 74, pp. 4187–4206.

Campbell, Joseph (1974): *The Mythic Image*, Princeton University Press.

Canfield, D.E., S.W. Poulton, and G.M. Narbonne (2007): "Late-Neoproterozoic deep-ocean oxygenation and the rise of animal life," *Science,* vol. 315, pp. 92–95.

Carte géologique de l'Europe, 1:500,000, Sheet D3 (2nd edition, 1966) © Bundesanstalt für Bodenforschung /Unesco.

Carte métallogénique de l'Europe, 1:2,500,000, Sheet 2 - Europe N (1968–1970) © UNESCO et Bureau de Recherches Géologiques et Minières.

Chatterjee, Sankar, Arghya Goswami, and Christopher R. Scotese (2013): "The longest voyage: Tectonic, magmatic, and paleoclimatic evolution of the Indian plate during its northward flight from Gondwana to Asia," *Gondwana Research*, vol. 23, no. 1, pp. 238–267.

Cheilletz, A., G. Feraud, G. Giuliani, and C.T. Rodriguez (1994): "Time-pressure and temperature constraints on the formation of Colombian emeralds; an $^{40}Ar/^{39}Ar$ laser microprobe and fluid inclusion study," *Economic Geology*, vol. 89, no. 2, pp. 361–380.

Chen, Jun-Yuan, D.J. Bottjer, P. Oliveri, S.Q. Dornbos, F. Gao, S. Ruffins, H.-M. Chi, C.-W. Li, and E.W. Davidson (2004): "Small bilaterian fossils from 40 to 55 million years before the Cambrian," *Science*, vol. 305, pp. 218–222.

Clark, Chris, Ian C.W. Fitzsimons, David Healy, and Simon L. Harley (Aug. 2011): "How does the Continental Crust Get Really Hot?" *Elements*, vol. 7, pp. 235–240.

Clausen, Anette (Oct. 2013): "Gemstones of Greenland," *33rd International Gemmological Conference IGC* (abstract proceedings) Hanoi, Vietnam, pp. 122–125. http://www.igc-gemmology.org.

Clery, Daniel (11 Oct. 2013): "Impact Theory Gets Whacked," *Science,* vol. 342, pp. 183–185.

Clites, Erica C., Mary L. Droser, and James G. Gehling (2012): "The advent of hard-part structural support among the Ediacara biota: Ediacaran harbinger of a Cambrian mode of body construction," *Geology,* vol. 40, no. 4, pp. 307–310.

Cohenour, Robert E., and Bryon J. Sharp (1968): "The Impact Theory: Asteroids and the Earth-Moon System," *GeoScience News*, vol. 1, no. 3, pp. 9–11, 32–34.

Commission for the Geological Map of the World: "Seismotectonic Map of the World – Five millennia of earthquakes around the world," *1:50,000,000* (1st edition, 2001), Abdolazim Haghipour, general coordinator.

Condie, Kent (1980): "Origin and early development of the earth's crust," *Precambrian Research*, vol. 11, pp. 183–197. DOI:10.1016/0301-9268(80)90064-9.

Condon, Edward U. (1969): *Scientific Study of Unidentified Flying Objects*, Bantam, New York.

Corliss, William (Jan. 2010): *Anomaly Register #9*, Glen Arm, Md.

Coste, Eugene (1905): "The volcanic origin of oil," *Trans. of the American Institute of Mining Engineers,* vol. 35, pp. 288–297.

Courtillot, Vincent E., and Paul R. Renne (2003): "On the ages of flood basalt events - Sur l'âge des trapps basaltiques," *C. R. Geoscience,* vol. 335, pp. 113–140. DOI: 10.1016/S1631-0713(03)00006-3.

Culver, S.J. (1991): "Early Cambrian foraminifera from West Africa," *Science,* vol. 254, pp. 689–691.

Daim, Wilfried (1958): *Der Mann, der Hitler die Ideen gab*, Isar Verlag, Munich.

Dalziel, Ian W.D. (Nov. 1992): "On the Organization of American Plates in the Neoproterozoic and the Breakout of Laurentia," *GSA Today*, vol. 2, no. 11.

Davies, Geoffrey F. (1992): "On the emergence of plate tectonics," *Geology*, vol. 20, no. 11, pp. 963–966.

Dawson, J.B. (1970). "The structural setting of African kimberlite magmatism," pp. 321–335 in T.N. Clifford and I.G. Gass, editors, *African Magmatism and Tectonics,* Oliver and Boyd Ltd., Edinburgh.

Derry, Louis A. (10 Mar. 2006): "Fungi, Weathering, and the Emergence of Animals," *Science*, vol. 311, pp. 1386–1387.

Deschamps Yves, Elizabeth Le Goff, J.P. Milesi, Cédric Simonet, S. Muhongo, B.A. Mcharo, H.A.B. Kampunzu, P. Pinna, F. Ralay, and John M. Saul (2006): "Gemstone Resources and potential of East Africa and Madagascar," *21ème colloque de géologie africaine*, Maputo, Mozambique.

Deschamps, Yves, Jean-Pierre Milesi, Sospeter Muhongo, Henri Kampunzu, Elizabeth Le Goff, Patrice Pinna, Cédric Simonet, Emmanuel Duguey, Alain Marot, Yves Guillou, Adrien Pouradier, Francis Ralay, Claude Heinry, Frédéric Chêne, Boniface Mcharo, and Christian Joannes (2004): *Géologie et Minéralisations À Gemmes D'Afrique de l'Est - Geology and Gemstone Deposits of East Africa, 1:4,000,000.* ISBN 2-7159-0944-6.

Des Marais, D.J., H. Strauss, R.E. Summons, and J.M. Hayes (1992): "Carbon isotope evidence for the stepwise oxidation of the Proterozoic environment," *Nature,* vol. 359, pp. 605–609.

Dietz, Robert S. (1961): "Astroblemes," *Scientific American*, vol. 205, pp. 50–58.

Dimitrov, Lyobomir I. (2003): "Mud volcanoes – a significant source of atmospheric methane," *Geo-Marine Letters,* vol. 23, no. 3/4, pp. 155–161. DOI: 10.1007/s00367-003-0140-3.

Dissanayake, C.B., and Rohana Chandrajith (1999): "Sri Lanka-Madagascar Gondwana Linkage: Evidence for a Pan-African Mineral Belt," *Jour. Geology*, vol. 107, pp. 223–235.

Dmitrievsky, Anatoly N., and Boris M. Valyaev (2013): "Degassing of the Earth and Genesis of Hydrocarbon Fluids and Fields," *Kazan workshop on abiotic hydrocarbons, 13–17 April 2013* (abstract) I.N. Plotnikova and A.S. Borisov, eds., Kazan Federal University, Kazan. ISBN 978-5-00019-053-1.

Dobzhansky, Theodosius (1973): "Nothing in biology makes sense except in the light of evolution," *American Biology Teacher*, vol. 35. pp. 125–129.

Dzik, J. (2005): "Behavioral and anatomical unity of the earliest burrowing animals and the cause of the 'Cambrian Explosion,'" *Paleobiology*, vol. 31, pp. 503–521.

Ehrenfreund. P., and J. Cami (2010): "Cosmic carbon chemistry: from the interstellar medium to the early Earth," *Cold Springs Harbor Perspectives in Biology*. DOI: 10.1101/cshperspect.a002097.

Ehrlich, Robert (2001): *Nine Crazy Ideas in Science*, Princeton University Press.

Ehrmann, Martin (c.1960?): unpublished text, available at http://www.palagems.com/ruby_mines_mogok3.htm (accessed 13 June 2013).

Eitel, Ernest John (1979): *Feng Shui*, Pentacle Books, Bristol (3rd ed.; first published 1873).

El Albani, Abderrazak, S. Bengtson, D.E. Canfield, A. Bekker, R. Macchiarelli, A. Mazurier, E.U. Hammarlund, P. Boulvais, J.-J. Dupuy, C. Fontaine, F.T. Fürsich, F. Gauthier-Lafaye, P. Janvier, E. Javaux, F. Ossa Ossa, A.-C. Pierson-Wickmann, A. Riboulleau, P. Sardini, D. Vachard, M. Whitehouse, and A. Meunie (1 July 2010): "Large colonial organisms with coordinated growth in oxygenated environments 2.1 Gyr ago," *Nature*, vol. 466, pp. 100–104. DOI: 10.1038/nature09166.

Encyclopaedia Britannica, 15th ed., "Sacred Kingship."

Endersbee, Lance (2005): *A Voyage of Discovery*. ISBN 0-646-45301-7.

Evans, Bernard W., Keiko Hattori, and Alain Baronnet (2013): "Serpentinites: What, Why, Where?" *Elements*, vol. 9, pp. 99–106.

Fassett, Caleb I., Seth J. Kadish, James W. Head III, Sean C. Solomon, Robert G. Strom (2011): "The global population of large craters on Mercury and comparison with the Moon," *Geophysical Research Letters*, vol. 38, L10202, 6 pp.

Fenchel, T., and B.J. Finlay (1995): *Ecology and Evolution in Anoxic Worlds*, 276 pp., Oxford University Press, Oxford, UK.

Feneyrol, Julien (2012): *Pétrologie, géochimie et genèse des gisements de tsavorite associés aux gneiss et roches calco-silicatées graphiteux de Lemshuku et Namalulu (Tanzanie)*, Ph.D. thesis, Institut National Polytechnique de Lorraine, Vandoeuvre-lès-Nancy, France, 617 pp.

Feneyrol, J., G. Giuliani, D. Ohnenstetter, E. Le Goff, C. Simonet, M. Rakotondrazafy, D. Ichang'i, E. Malisa, Y. Dechamps, A.E. Fallick, T. Kahn, and C. Nyamai (2013): "Les gemmes néoprotérozoïques de la ceinture métamorphique mozambicaine du supercontinent Gondwana (1000–542 Ma)," *Règne Minéral*, Cahier 2 - 2013, pp. 13–47.

Feneyrol, J., G. Giuliani, D. Ohnenstetter, E.P. Malisa, A.F.M. Rakotondrazafy, D. Ichang'i, E. Venance, N. Voary, M. Ranatsenho, E. Omito, and C. Nyami (2011): "La tsavorite dans la ceinture métamorphique néoprotérozoic Mozambicaine: un exemple de marqueur géologique entre l'Afrique de l'Est et Madagascar," *Géologie et Métallogénie de Madagascar*, Société géologique de France, Paris (Dec. 6–7), www.google.com/search?q=CEINTURE+METAMORPHIQUE+NEOPROTEROZOIQUE+MOZAMBICAINE&ie=utf-8&oe=utf-8&aq=t&rls=org.mozilla:fr:official&client=firefox-a (1 Jan. 2012).

Ferrière, Ludovic (2008): *Shock metamorphism and geochemistry of impactites from the Bosumtwi impact structure: A case study of shock-induced deformations and transformations in quartz and associated methodology*, Ph.D. thesis, Petrography-Geochemistry, University of Vienna, Austria.

Ferrière, Ludovic, Christian Koeberl, and Wolf Uwe Reimold (2009): "Characterisation of ballen quartz and cristobalite in impact breccias: new observations and constraints on ballen formation," *European Jour. Mineralogy*, 15 pp. DOI: 10.1127/0935-1221/2009/0021-1898.

Ferrière, Ludovic, and Gordon R. Osinski (2010): "Shatter cones and associated shock-induced microdeformations in minerals – New investigations and implications for their formation," *41st Lunar and Planetary Science Conference*, Poster and Abstract #1392, Houston.

Fitzsimons, I.C.W., and B. Hulscher (June 2005): "Out of Africa: Detrital zircon provenance of central Madagascar and Neoproterozoic terrane transfer across the Mozambique Ocean," *Terra Nova*, vol. 17, no. 3, pp. 224–235. DOI: 10.1111/j.1365-3121.2005.00595.x.

French, Bevan M. (1998): "Traces of catastrophe: A handbook of shock-metamorphic effects," In: *Terrestrial Meteorite Impact Structures*, Lunar and Planetary Institute, Houston, Contribution no. 954; 120 pp. http://www.lpi.usra.edu/publications/books/CB-954/CB-954.intro.html.

Fritz, Harald, Veronika Tenczer, Christoph Hauzenberger, Eckart Wallbrecher, and Muhongo Sospeter (2009): "Hot granulite nappes - tectonic styles and thermal evolution of the Proterozoic Granulite Belt in East Africa, *Tectonophysics*, vol. 477, pp. 160–173.

Gahalaut, Vineet K., Bhaskar Kundu, Sunil Singh Laishram, Joshi Catherine, Arun Kumar, M. Devchandra Singh, R.P. Tiwari, R.K. Chadha, S.K. Samanta, A. Ambikapathy, P. Mahesh, Amit Bansal, and M. Narsaiah (2013): "Aseismic plate boundary in the Indo-Burmese wedge, northwest Sunda Arc," *Geology*, vol. 41, pp. 235–238. DOI: 10.1130/G33771.1.

Gallant, René (1964): *Bombarded Earth*, John Barker, London.

Garde, Adam A., Iain McDonald, Brendan Dyck, and Nynke Keulen (2012): "Searching for giant, ancient impact structures on Earth: The Mesoarchaean Maniitsoq structure, West Greenland," *Earth and Planetary Science Letters*, vol. 337–338, pp. 197–210. DOI: 10.1016/j.epsl.2012.04.026.

Garnier, Virginie (2003): *Les gisements de rubis associés aux marbres de l'Asie Centrale et du Sud-Est: genèse et caractérisation isotopique*, Ph.D. thesis, Institut National Polytechnique de Lorraine, Vandoeuvre-lès-Nancy, 373 pp.

Garnier V., G. Giuliani, D. Ohnenstetter, A.E. Fallick, J. Dubessy, D. Banks, V. Hoàng Quang, Th. Lhomme, H. Maluski, A. Pêcher, K.A. Bakhsh, L. Pham Van, T. Phan Trong, and D. Schwarz (2008): "Marble-hosted ruby deposits from central and Southeast Asia: towards a new genetic model," *Ore Geology Rev.*, vol. 34, pp. 169–191.

Garnier, V., H. Maluski, G. Giuliani, D. Ohnenstetter, and D. Schwarz (2006): "Ar-Ar and U-Pb ages of marble-hosted ruby deposits from central and southeast Asia," *Canadian Jour. Earth Science*, vol. 43, pp. 509–532.

Garnier, V., D. Ohnenstetter, G. Giuliani, A.E. Fallick, T. Phang Trong, V. Hoang Quang, L. Pham Van, and D. Schwarz (2005): "Basalt petrology, zircon ages and sapphire genesis from Dak Nong, Southern Vietnam," *Mineralogical Magazine*, vol. 69, no. 1, pp. 21–38.

Geological World Atlas, 1:10,000,000, sheets 2, 3, and 9; © C.G.M.W.-Unesco.

Ginzburg, A.I. (1984): "The geological condition of the location and the formation of granitic pegmatites," *Proc. 27th International Geological Congress*, Moscow, vol. 15, pp. 245–259.

Giuliani, Gaston (2010): "Les gisements de tsavorite associés aux gneiss graphiteux de Tanzanie, Madagascar et du Kenya: genèse et caractérisation isotopique," Rapport Projet CESSUR, AO INSU 2010, pp. 1–10.

Giuliani, Gaston, Anthony Fallick, Daniel Ohnenstetter, and Guy Pegere (2009): "Oxygen isotopes composition of sapphires from the French Massif Central: implications for the origin of gem corundum in basaltic fields," *Mineral Deposita*, vol. 44, pp. 221–231.

Giuliani, Gaston, Anthony Fallick, Michel Rakotondrazafy, Daniel Ohnenstetter, Alfred Andriamamonjy, Théogène Ralantoarison, Saholy Rakotosamizanany, Marie Madison, Yohann Offant, Virginie Garnier, Christian Dunaigre, Dietmar Schwarz, Alain Mercier, Vohangy Ratrimo, and Bruno Ralison (2007): "Oxygen isotope systematics of gem corundum deposits in Madagascar: Relevance for their geological origin," *Mineral Deposita*, vol. 42, pp. 251–270.

Giuliani G., D. Ohnenstetter, A.E. Fallick, and J. Feneyrol (2011): "State of the art in the formation of high-value colored gemstones," *Gems & Gemology*, vol. 47, pp. 108–110.

Giuliani, Gaston, Daniel Ohnenstetter, and Julien Feneyrol (2013): "La Reine Bleue de l'Afrique de l'Est: la Tanzanite, une zoïsite vanadifère," *Règne Minéral*, Cahier 2 - 2013, pp. 71–81.

Glukhovskiy, M.Z. (1978): "Ring structures and linear faults in the Aldan Shield and Stanovoy Region," *Geotectonics*, vol. 10, p. 331.

Goddard, Alison (15 Apr. 1995): "'Burst zip' undermines theories of ocean birth," *New Scientist*, p. 14.

Godwin, Colin I. (1973): "Shock brecciation, an unrecognized mechanism for breccia formation in the porphyry environment?" *Proc., Geol. Association of Canada*, vol. 25, pp. 9–12.

Gold, Thomas (26 Mar. 1955): "Instability of the Earth's axis of rotation," *Nature*, vol. 175, no. 4456, pp. 526–529.

Gold, Thomas (Dec. 1984): "The origin of natural gas and petroleum, and the prognosis for future supplies," manuscript prepared for publication in *Annual Review of Energy*, 20 pp.

Gold, Thomas (1987): *Power from the Earth: Deep Earth gas – energy for the future*, J.M. Dent & Sons Ltd., London.

Gold, Thomas (1993): "The origin of methane in the crust of the Earth," *U.S. Geological Survey Professional Paper*, 1570, pp. 57–80.

Gold, Thomas (1999): *The Deep Hot Biosphere: the myth of fossil fuel*, Springer, New York.

Gold, Thomas [edited by Simon Mitton] (2012): *Taking the Back off the Watch: A Personal Memoir,* Astrophysics and Space Science Library 381, Springer.

Gold, Thomas, and Steven Soter (Dec. 1979): "The Deep Earth Gas Hypothesis," Cornell University Center for Radiophysics and Space Research, *CRSR* 736, 39 pp.

Gold, Thomas, and Steven Soter (1982): "Abiogenic methane and the origin of petroleum," *Energy Exploration & Exploitation,* vol. 1, no. 2, pp. 89–104.

Gold, Thomas, and Steven Soter (1984/85): "Fluid ascent through the solid lithosphere and its relation to earthquakes," *Pure and Applied Geophysics*, vol. 122, no. 2/4, pp. 492–530.

Gorlenko, V.M., S.I. Zhmur, V.I. Duda, N.E. Suzina, G.A. Osipov, and V.V. Dmitriev (2000): "Fine structure of fossilized bacteria in Volyn kerite," *Origins of Life and Evolution of the Biosphere,* vol. 30, pp. 567–577.

Gould, Stephen Jay (1991): *Bully for Brontosaurus,* W.W. Norton and Company.

Gould, Stephen Jay (Dec. 1994): "Common pathways to illumination," *Natural History,* vol. 103, no. 12, pp. 10–20.

Graham, I.T., F.L. Sutherland, G. Webb, and C.M. Fanning (2004): "Polygenetic corundums from New South Wales gemfields, Australia," In: Alexander I. Khanchuk *et al.*, eds., *Metallogeny of the Pacific Northwest: Tectonics, Magmatism and Metallogeny of Active Continental Margins*, pp. 336–339, Dalnauka, Vladivostok.

Graham, Ian T., Lin Sutherland, Khin Zaw, Victor Nechaev, and Alexander Khanchuk (2008): "Advances in our understanding of the gem corundum deposits of the West Pacific continental margins intraplate basaltic fields," *Ore Geology Rev.*, vol. 34, pp. 200–215.

Graham, James (1992): *Cancer Selection: The New Theory of Evolution*, 213 pp., Aculeus Press, Lexington, Va.

Grantham, G.H., P.H. Macey, B.A. Ingram, M.P. Roberts, R.A. Armstrong, T. Hokada, K. Shiraishi, C. Jackson, A. Bisnath, and V. Manhica (2008): "Terrane correlation between Antarctica, Mozambique and Sri Lanka; comparisons of geochronology, lithology, structure and metamorphism and possible implications for the geology of southern Africa and Antarctica," *Special Publications, Geological Society*, London, vol. 308, pp. 91–119.

Green, T.H., S.Y. Wass, and J. Ferguson (1978): "Experimental study of corundum solubility in Basalts," Abstract Program, *3rd Australian Geological Convention*, Townsville, Australia, p. 34.

Gribble, Gordon W. (March-April 2001): Letter to the Editors, *Scientific American*, p. 407.

Gribble, Gordon W., and others (Jan. 1995): exchange of correspondence, *Environmental Science & Technology*, vol. 29, no. 1, p. 8A. DOI: 10.1021/es00001a705.

Grieve, Richard A.F. (1991): "Terrestrial impact: The record of the rocks," *Meteoritics*, vol. 26, pp. 175–194.

Grieve, Richard A.F., and M.J. Cintala (1992): "An analysis of differential impact melt-crater scaling and implications for the terrestrial impact record," *Meteoritics*, vol. 27, pp. 526–238.

Grunow, Anne, Richard Hanson, and Terry Wilson (1996): "Were aspects of the Pan-African deformation linked to Iapetus opening?" *Geology*, vol. 24, no. 12, pp. 1063–1066.

Guillot, Stéphane, and Keiko Hattori (2013): "Serpentinites: Essential roles in geodynamics, arc volcanism, sustainable development and the origin of life," *Elements*, vol. 9, pp. 95–98.

Guo J., S.Y. O'Reilly, and W.L. Griffin (1996): "Corundum from basaltic terrains: A mineral inclusion approach to the enigma," *Contributions to Mineralogy and Petrology*, vol. 22, pp. 368–386.

Haapala, Ilmari, and O. Tapani Rämö (1992): "Tectonic setting and origin of the Proterozoic rapakivi granites of southeastern Fennoscandia," *Transactions of the Royal Society of Edinburgh Earth Sciences*, vol. 83, nos. 1–2, pp. 165–171. DOI: 10.1017/S0263593300007859

Hagar, Stansbury (1900): "The Celestial Bear," *Journal of American Folk-lore*, vol. XIII, no. 49, pp. 92–103.

Haghipour: see Commission for the Geological Map of the World.

Hansen, Vicki L. (2007): "Subduction origin on early Earth: A hypothesis," *Geology*, vol. 35, no. 12, pp. 1059–1062. DOI: 10.1130/G24202A.1.

Harder, B. (8 Nov. 2003): "Calcium makes germs cluster: ion dilution leads cholera bacteria to disperse," *Science News*, vol. 164, no. 19, pp. 293–294.

Harris, Bradford (Nov.–Dec. 2013): "Evolution's other narrative: Why science would benefit from a symbiosis-driven history of speciation," *American Scientist*, vol. 101, no. 6, pp. 410–413.

Harshbarger, J.C., and D.I. Gibson (1982): "Ganglioneuroblastoma in a trematode, *Otodistomum plunketi* Fyfe, 1953," *Invertebrate Pathology and Microbial Control Proceedings. Third International Colloquium on Invertebrate Pathology*, University of Sussex, Brighton, UK, pp. 280–285.

Hartmann, W.K. (2010): "Craters, first significant figures, and the pivotal year of 4,000,000,000 B.C.E.," *Meteoritics & Planetary Science*, vol. 45 Supplement, Abstract 5052, p. A77.

Hartmann, W.K., and D.R. Davis (1975): "Satellite-sized planetesimals and lunar origin," *Icarus* vol. 24, no. 4, pp. 504–514.

Hasan, N.M., G.E. Adams, M.C. Joiner, J.F. Marshall, and I.R. Hart (1998): "Hypoxia facilitates tumor cell detachment by reducing expression of surface adhesion molecules and adhesion to extracellular matrices without loss of cell viability," *British Journal of Cancer*, vol. 77, pp. 1799–1805.

Hazen, Robert M., Robert T. Downs, Adrian P. Jones, and Linda Kah (2013): "Carbon mineralogy and crystal chemistry," In: *Carbon in Earth,* R.M. Hazen, A.P. Jones, and J.A. Baross, eds., *Reviews in Mineralogy and Chemistry*, vol. 75, pp. 7–46, Mineralogical Society or America - Geochemical Society.

Head, James W. III, Caleb I. Fassett, Seth J. Kadish, David E. Smith, Maria T. Zuber, Gregory A. Neumann, and Erwan Mazarico (17 Sept. 2010): "Global distribution of large lunar craters: Implications for resurfacing and impactor populations," *Science*, vol. 329, no. 5998, pp. 1504–1507. DOI: 10.1126/science.1195050.

Hedberg, Hollis D. (1964): "Geological aspects of petroleum," *AAPG Bulletin*, vol. 48, no. 11, pp. 1755–1803.

Heinemann, S., H. Ehrlich, C. Knieb, and T. Hanke (2007): "Biomimetically inspired hybrid materials based on silicified collagen," *International Jour. Materials Research* vol. 98, pp. 1–7.

Hindley, Keith (11 August 1977): "Meteorites at Cambridge," *New Scientist*, p. 353.

Hindley, Keith (20 Mar. 1980): "The riches in the ghost rings," *The Guardian,* p. 13.

Hinrichs, K.-U., J.M. Hayes, W. Bach, A. Spivack, L.R. Hmelo, N. Holm, C.G. Johnson, and S.P. Sylva (2006): "Biological formation of ethane and propane in the deep marine subsurface," *Proceedings of the National Academy of Sciences, U.S.A.*, vol. 103, pp. 14684–14689. DOI:10.1073/pnas.0606535103.

Hitchcock, Charles H. (1865): "The Albert Coal, or albertite, of New Brunswick," *American Journal of Science,* vol. 2, pp. 267–273.

Hobohm, U. (14 Feb. 2005): "Fever therapy revisited," *British Journal of Cancer*, vol. 92, no. 3, pp. 421–425. http://www.nature.com/bjc/journal/v92/n3/full/6602386a.html.

Hoffman, Paul F. (1988): "United plates of America, The birth of a craton: Early Proterozoic assembly and growth of Laurentia," *Ann. Rev. Earth and Planetary Sciences,* vol. 16, pp. 543–603. DOI: 10.1146/annurev.ea.16.050188.002551.

Holland, H.D. (2002): "Volcanic gases, black smokers, and the great oxidation event," *Geochimica et Cosmochimica Acta*, vol. 66, pp. 3811–3826.

Hollis J.D., F.L. Sutherland, and R.E. Pogson (1983): "High pressure minerals and the origin of the Tertiary breccia pipe, Ballogie Gem Mine, near Proston, Queensland," *Rec., Australian Museum*, vol. 35, pp. 181–194.

Hoorn, C[arina], F.P. Wesselingh, H. ter Steege, M.A. Bermudez, A. Mora, J. Sevink, I. Sanmartín, A. Sanchez-Meseguer, C.L. Anderson, J.P. Figueiredo, C. Jaramillo, D. Riff , F.R. Negri, H. Hooghiemstra, J. Lundberg, T. Stadler, T. Särkinen, A. Antonelli (12 Nov. 2010): "Amazonia through time: Andean uplift, climate change, landscape evolution and biodiversity," *Science*, vol. 330, pp. 927–931.

Hoşgörmez, H., G. Etiope, and M.N. Yalçın (2008): "New evidence for a mixed inorganic and organic origin of the Olympic Chimaera fire (Turkey): A large onshore seepage of abiogenic gas," *Geofluids*, vol. 8; pp. 263–273. DOI: 10.1111/j.1468-8123.2008.00226.x.

Howard, J.M., and W.D. Hanson (2008): "Geology of the Crater of Diamonds State Park, Pike County, Arkansas," *Arkansas Geological Survey, State Park Series* 03, Little Rock, Ark. http://www.geology.ar.gov/pdf/crater_diamond__sps_06_%2024_8.pdf.

Hu WenXuan, Song YuCai, Chen XiaoMing, Toa MingXin, and Zhang LuiPing (2007): "Noble gases in corundum megacrysts from basalts in Changle, Shandong Province, eastern China," *Revue, Chinese Science Bull.*, vol. 52, no. 3, pp. 380–387. ISSN 1001-6538. DOI: 10.1007/s11434-007-0044-0.

Huldtgren, Therese, John A. Cunningham, Chongyu Yin, Marco Stampanoni, Federica Marone, Philip C.J. Donoghue, and Stefan Bengtson (23 Dec. 2011): "Fossilized nuclei and germination structures identify Ediacaran 'animal embryos' as encysting protists," *Science*, vol. 334, no. 6063, pp. 1696–1699. DOI: 10.1126/science.1209537.

Hureau, Jean-Claude, and Denis Escudier (2005): "Théodore Monod: 90 années de publications," Publications Scientifiques du Muséum national d'histoire naturelle, Paris.

Hyrsl, Jaroslav (2011): "Historic use of olivine – the origin of peridots in baroque-period jewellery," *Abstract Proceedings, 32nd International Gemmological Conference*, Interlaken, Switzerland, pp. 108–109.
I-Ming Chou (2010): "Two types of optical cells for the study of C-H-O-S fluids at Moho and sedimentary basins," http://sess.pku.edu.cn/organization/gwgy/EN/news-zhouyiming.html.
Isachsen, Yngvar W. (1978): "Large circles on the Earth's surface," Letter to the Editors, *Nature*, vol. 276, no. 5687, p. 535.
Iyer, L.A.N. (1953): "The geology and gem-stones of the Mogok Stone Tract, Burma," *Memoirs, Geological Survey of India*, vol. 82.
Izokh, A.E., S. Smirnov, V.V. Egorova, Tran Tuan Anh, S.V. Kovyazin, Ngo Thi Phuong, and V.V. Kalinina (2010): "The condition of formation of sapphire and zircon in the areas of alkali-basaltoid volcanism in Central Vietnam," *Russian Geology and Geophysics*, vol. 51, pp. 719–733. DOI: 10.1016/j.rgg.2010.06.001.
Jackson, N.J., and C.R. Ramsay (1980): "What is the 'Pan-African'? A consensus is needed," *Geology*, vol. 8, no. 5, pp. 210–211.
Jacob, François (10 June 1977): "Evolution and tinkering," *Science*, vol. 196, pp.1161–1166.
Jahn, B.M., R. Caby, and P. Monie (2001): "The oldest UHP eclogites of the world: Age of UHP metamorphism, nature of protoliths and tectonic implications," *Chemical Geology*, vol. 178, pp. 143–158.
Jastrow, Morris Jr.: (1911): *Aspects of Religious Belief and Practice in Babylonia and Assyria*, American Lectures on the History of Religions, G.P. Putnam's Sons, New York.
Jerome, S.E., and D.R. Cook (1967): "Relation of some metal mining districts in the western United States to regional tectonic environments and igneous activity," *Bull. Nevada Bureau of Mines*, vol. 67.
Johnson, B.C., and H. Jay Melosh (2012): "Impact spherules as a record of an ancient heavy bombardment of Earth," *Nature*, vol. 485, pp. 75–78. DOI: 10.1038/nature10982.
Jones, Alan G. (1987): "Are impact-generated lower-crustal faults observable?" *Earth and Planetary Science Letters*, vol. 85, nos. 1–3, pp. 248–252.
Jones, Tom B., ed. (1969): *The Sumerian Problem*, John Wiley & Sons, New York.
Julian, W., W. Lounsberry, A. Tamura, R. Van Loenen, and M.P. Wright (1962 [1963]): *Meia Meia, Quarter Degree Sheet 143, 1:125,000*, Geol. Survey Tanganyika, Dodoma, Tanzania.
Kane, Robert E., and Robert C. Kammerling (1992): "Status of Ruby and Sapphire Mining in the Mogok Stone Tract," *Gems & Gemology*, vol. 28, no. 3, pp. 152–174.
Kang, Kyung-Sun, and James E. Trosko (2011): "Stem cells in toxicology: Fundamental biology and practical considerations," *Toxicological Sciences*, vol. 120(S1), pp. S269-S289. DOI: 10.1093/toxsci/kfq370.
Kavanagh, Janine L., and R. Stephen J. Sparks, (2009): "Temperature changes in ascending kimberlite magma," *Earth and Planetary Science Letters*, vol. 286, pp. 404–413.
Kelly, Allan O., and Frank Dachille (1953): *Target: Earth; the role of large meteors in earth science*, Pensacola Engraving Co., Pensacola, Fla.
Kenya Mines and Geological Department, *Ann. Report 1953*.
Kerr, Richard A. (27 Nov. 1987): Research News, *Science*, vol. 238, p. 1227.
Kerr, Richard A. (8 Jan. 1988): Research News, *Science*, vol. 239, p. 145.
Kerr, Richard A. (2 Sept. 1994): "The Solar System's new diversity," *Science*, vol. 265, no. 5177, pp. 1360–1362.
Kierek, K., and P.I. Watnick (2003): "Environmental determinants of *Vibrio cholerae* biofilm development," *Applied and Environmental Microbiology*, vol. 69, pp. 5079–5088.
Kievlenko, Eugenii Ya. (2003): *Geology of Gems*, Ocean Pictures Ltd., Littleton, Colo.
King, Nicole, Christopher T. Hittinger, and Sean B. Carroll (18 July 2003): "Evolution of key cell signaling and adhesion protein families predates animal origins," *Science*, vol. 301, pp. 361–363.

Kirschvink J.L., and J.W. Hagadorn (2000): "A grand unified theory of biomineralization," Ch. 10, pp.139–150, In: E. Bäuerlein, ed., *The Biomineralisation of Nano- and Micro-Structures*, Wiley-VCH Verlag GmbH, Weinheim, Germany. http://www.roaringlionpublishing.com/tony_uploads/Unified_Theory_Biomineralization.pdf.

Kirschvink, Joseph L., Robert L. Ripperdan, and David A. Evans (25 July 1997): "Evidence for a large-scale reorganization of early Cambrian continental masses by inertial interchange true polar wander," *Science*, vol. 277, pp. 541–545. DOI: 10.1126/science.277.5325.541.

Knoll, A.H., and J.H. Lipps (1993): "Evolutionary history of prokaryotes and protests," In: J.H. Lipps, ed., *Fossil Prokaryotes and Protists*, pp. 19–29, Blackwell Scientific, Boston, Mass.

Knoll, A.H., M.R. Walter, G.M. Narbonne, and N. Christie-Blick (2006): "The Ediacaran period: A new addition to the geologic time scale," *Lethaia*, vol. 39, pp. 13–30.

Kochemasov, Gennady G. (1983): "The Congo Craton: An old impact structure?" *14th Lunar and Planetary Science Conference*, pp. 377–378 (abstract) Houston.

Konopliv, A.S., S.W. Asmar, E. Carranza, W.L. Sjogren, and D.N. Yuan (2001): "Recent gravity models as a result of the Lunar Prospector Mission," *Icarus*, vol. 150, pp. 1–18. DOI: 10.1006/icar.2000.6573 http://science1.nasa.gov/science-news/science-at-nasa/2006/06nov_loworbit/.

Kriegsman, L.M. (1994): "Evidence for a fold nappe in the high-grade basement of central Sri Lanka: terrane assembly in Pan-African lower crust?" In: M. Raith, and S. Hoernes, eds., Tectonic, Metamorphic and Isotopic Evolution of Deep Crustal Rocks, with Special Emphasis on Sri Lanka, *Precambrian Research*, vol. 66, pp. 59–76.

Kropotkin, P.N. (1995): "On the history of science: Professor N.A. Koudryavtsev (1893–1971) and the development of the theory of origin of oil and gas," *Bull. Moscow Soc. Nature Res.*, vol. 70, no. 1.

Kropotkin, P.N., and B.M. Valiaev (1976): "Development of a theory of deep-seated (inorganic and mixed) origin of hydrocarbons," In: *Goryuchie Islopaemye: Problemy Geologii i Geokhimii Naftidov i Bituminoznyky Porod [Combustible Mineral Resources. Problems of Geology and Geochemistry of Petroleum Bitumen and Bituminous Rocks]*, N.B. Vassoevich *et al.*, eds., Akad. Nauk SSSR, pp. 133–144.

Kudryavtsev N.A. (1959): "Geological proof of the deep origin of petroleum," *Trudy Vsesoyuz. Neftyan. Nauch. Issledovatel Geologoraz Vedoch. Inst.*, no. 132, pp. 242–262 (undated Associated Technical Services translation).

Kump, L.R., and M.E. Barley (2007): "Increased subaerial volcanism and the rise of atmospheric oxygen 2.5 billion years ago," *Nature*, vol. 448, pp. 1033–1036.

Kutcherov V.G., N.A. Bendeliani, V.A. Alekseev, and J. F. Kenney (2002): "Synthesis of hydrocarbons from minerals at pressures up to 5 GPa," *Doklady Physical Chemistry*, vol. 387, nos. 4–6, pp. 328–330. Translated from *Doklady Akademii Nauk*, vol. 387, no. 6 (2002), pp. 789–792.

Kutcherov, Vladimir G., and Vladilen A. Krayushkin (2010): "Deep-seated abiogenic origin of petroleum: from geological assessment to physical theory," *Reviews of Geophysics*, vol. 48, pp. 1–30.

Lahti, Seppo I., and Kari A. Kinnunen (1993): "A new gem beryl locality, Luumäki, Finland," *Gems & Gemology*, vol. 29, no. 1, pp. 30–37.

Lanz von Liebenfels, Jörg [Adolf Josef Lanz] (1931): *Bibliomystikon oder Die Geheimbibel der Eingeweihten*: vol.3: Theozoikon, Privatdruckverlag "Bibliomystikon," Verlag H. Reichstein, Pressbaum bei Wien.

Laurs, B.M., J.C. Zwaan, C.M. Breeding, W.B. Simmons, D. Beaton, K.F. Rijsdijk, R. Befi, and A.U. Falster, (2008): "Copper-bearing (Paraíba-type) tourmaline from Mozambique", *Gems & Gemology*, vol. 44, no. 1, pp. 4–30.

Le Dain, Anne Yvonne, Paul Tapponnier, and Peter Molnar (1984): "Active faulting and tectonics of Burma and surrounding regions," *Jour. Geophysical Research*, vol. 89, no. B1, pp. 453–472.

Le Goff, Elizabeth, Yves Deschamps, A. Cocherie, C. Guerrot, and D. Ketto (2008): "Structural, petrological and geochronological constraints of the Tanzanian ruby belt," *Résumés, 22ème RST,* Nancy (Apr. 21–24, 2008), p. 133.

Le Goff, Elizabeth, Yves Deschamps, A. Cocherie, C. Guerrot, C. Lerouge, J.P. Milesi, P. Pinna, D. Ketto, E. Malisa, and S. Muhongo (2006): "Structural and petrological modelisation of the Tanzanian 'Ruby Belt'," *21ème Colloque de géologie africaine*, Maputo, Mozambique.

Leloup P.H., N. Arnaud, R. Lacassin, J.R. Kienast, T.M. Harrison, P.T. Trinh, A. Replumaz, and P. Tapponnier (2001): "New constrains on the structure, thermochronology and timing of the Ailao Shan - Red River shear zone," *Jour. Geophysical Research*, vol. 106, pp. 6683–6732.

Levin, M., and M. Mercola (1 Nov. 1998): "Gap junctions are involved in the early generation of left-right asymmetry," *Developmental Biology,* vol. 203, no. 1, pp. 90–105.

Levinson, Alfred A., John J. Gurney, and Melissa B, Kirkley (1992): "Diamond sources and production: Past, present and future", *Gems & Gemology*, vol. 28, no. 4, pp. 234–254.

Li Jianhua, and Shen Xuhui (2001): "Syntaxis faults at the northeastern corner of the Tibetan Plateau and the annular tectonics in southern Ordos," *Journal of Seismology and Geology*, vol. 23, no.1. http://d.wanfangdata.com.cn/Periodical_dzdz200101015.aspx.

Li, Z.X., D.A.D. Evans, and S. Zhang (2004): "A 90° spin on Rodinia: Possible causal links between the Neoproterozoic supercontinent, superplume, true polar wander and low-latitude glaciation," *Earth and Planetary Science Letters*, vol. 220, nos. 3–4, pp. 409–421.

Ligier-Belair, Gérard, and Guillaume Polidor (2009): *Voyage au Coeur d'une bulle de champagne*, Odile Jacob, Paris. ISBN 9782738123534.

Limtrakun, P., Zaw Khin, C.G. Ryan, and T.P. Mernagh (2001): "Formation of the Denchai gem sapphires, northern Thailand: evidence from mineral chemistry and fluid/melt inclusion characteristics," *Mineralogical Magazine*, vol. 65, pp. 725–735.

Lipps, Jere H., Thorsten Stoeck, and Micah Dunthorn (2012): "Fossil Tintinnids," In: *The Biology and Ecology of Tintinnid Ciliates: Models for Marine Plankton*, Ch. 8, First Edition, John R. Dolan, David J.S. Montagnes, Sabine Agatha, D. Wayne Coats, and Diane Stoecker, eds., John Wiley & Sons, Ltd.

Liu, ChiChing, Alan T. Linde, and Selwyn Sacks (2009): "Slow earthquakes triggered by typhoons," *Nature,* vol. 459, pp. 833–836. DOI: 10.1038/nature08202.

Lowenstein, W.R. (1979): "Junctional intercellular communication and the control of growth," *Biochimica et Biophysica Acta,* vol. 560, pp. 1–65.

Lowman, Paul D., Jr. (1976): "Crustal evolution in silicate planets: Implications for the origin of continents," *Jour. Geology*, vol. 84, no. 1, pp. 1–26.

Lowman, Paul D., Jr. (2002): *Exploring Space, Exploring Earth: New Understanding of the Earth from Space Research,* Cambridge University Press. ISBN: 9780521890625.

Lukjanova, V.T., and R.V. Lovzova (1994): "Carbon fibers in nature," *Carbon,* vol. 32, pp. 777–783.

Lyckberg, Peter, Vsevolod Chornousenko, and Wendell Wilson (2009): "Famous mineral localities: Volodarsk-Volynski, Zhitomir Oblast, Ukraine," *Mineralogical Rec.*, vol. 40, no. 6, pp. 473–506.

Mahadevan, L., R. Bendick, and Haiyi Liang (2010): "Why subduction zones are curved," *Tectonics,* vol. 29, TC6002, 10 pp.

Malisa, E.P. (1998): "Application of graphite as a geothermometer in hydrothermaly altered metamorphic rocks of the Merelani-Lelatema area, Mozambique Belt, Northeastern Tanzania," *Jour. African Earth Sciences*, vol. 26, no. 2, pp. 313–316.

Malisa, E., K. Kinnunen, and T. Koljonen, (1986): "Notes on fluid inclusions in vanadiferous zoisite (tanzanite) of the Merelani area, Tanzania," *Bull. Geol. Soc. Finland*, vol. 58, pp. 53–58.

Mallick, Brian C., and Dennis S. Hodge (10 Nov. 1981): "Adirondack radioisotope distribution and thermal models of the Adirondack Uplift," *EOS, Trans. American Geophysical Union*, vol. 62, no. 45, Abstract T2-1-A-13.

Maloof, Adam C., Catherine V. Rose, Robert Beach, Bradley M. Samuels, Claire C. Calmet, Douglas H. Erwin, Gerald R. Poirier, Nan Yao, and Frederik J. Simons (2010): "Possible animal-body fossils in pre-Marinoan limestones from South Australia," *Nature Geosciences*, vol. 3, pp. 653–659. DOI: 10.1038/ngeo934.

Mardirosian, Charles A. (1973): *Map of Mining districts and mineral deposits of Arizona (exclusive of oil and gas), 1:1,000,000*, Mineral Research Co., Albuquerque, NM.

Marty, Bernard (2012): "The origins and concentrations of water, carbon, nitrogen and noble gases on Earth," *Earth and Planetary Science Letters*, vols. 313–314, pp. 56–66.

Maruyama, S., and J.G. Liou (1998): "Initiation of UHP metamorphism and its significance on the Proterozoic/Phanerozoic boundary," *Island Arc*, vol. 7, pp. 6–35.

Matthews, Robert (17 Jan. 1999), *Sunday Telegraph* (London).

Matz, Mikhail V., Tamara M. Frank, N. Justin Marshall, Edith A. Widder, and Sönke Johnsen (9 Dec. 2008): "Giant deep-sea protist produces bilaterian-like traces," *Current Biology*, vol. 18, pp. 1–6. DOI: 10.1016/j.cub.2008.10.028.

Mayne, Rhiannon (Fall 2013): "The impact history of asteroidal meteorites," *Meteorite*, vol. 19, no. 3 pp. 26–28.

McConnell, Richard B. (1974): "Evolution of taphrogenic lineaments in continental platforms," *Geologisches Rundschau*, vol. 63, pp. 389–430.

Meert, Joseph G. (2003): "A synopsis of events related to the assembly of eastern Gondwana," *Tectonophysics*, vol. 362, pp. 1–40.

Meert, Joseph G., Rob van der Voo, and Samwel Ayub (1995): "Paleomagnetic investigation of the Neoproterozoic Gagwe lavas and Mbozi complex, Tanzania and the assembly of Gondwana," *Precambrian Research*, vol. 74, pp. 225–244.

Melosh, H. Jay (1975): "Mascons and the moon's orientation," *Earth and Planetary Science Letters*, vol. 25, no. 3, pp. 322–326.

Melosh, H. Jay (1982): "A schematic model of crater modification by gravity," *Jour. Geophysical Research*, vol. 87, no. B1, pp. 371–380. DOI: 10.1029/JB087iB01p00371.

Melosh, H. Jay (1989): *Impact Cratering: a Geologic Process,* Oxford Monographs on Geology and Geophysics, Series No. 11, Clarendon Press, Oxford.

Melosh, H. Jay (2000): "Can impacts induce volcanic eruptions?" *Catastrophic Events & Mass Extinctions: impacts and beyond, Lunar and Planetary Institute, LPI Contribution* 1053, Vienna (July 9–12, 2000), pp. 141–142.

Melosh, H. Jay, Andrew M. Freed, Brandon C. Johnson, David M. Blair, Jeffrey C. Andrews-Hanna, Gregory A. Neumann, Roger J. Phillips, David E. Smith, Sean C. Solomon, Mark A. Wieczorek, and Maria T. Zuber (28 June 2013): "The origin of lunar mascon basins," *Science*, vol. 340, no. 6140, pp. 1552–1555. DOI: 10.1126/science.1235768.

Melton, C.E., C.A. Salotti, and A.A. Giardini (1972): "The observation of nitrogen, water, carbon dioxide, methane, and argon as impurities in natural diamonds," *American Mineralogist*, vol. 57, pp. 1518–1523.

Mendeleev, D.L. (1877): "L'Origine du pétrole," *Revue Scientifique*, 2e Ser., VIII, pp. 409–416.

Mercier, Alain, P. Debat, and John M. Saul (1999): Exotic origin of the ruby deposits of the Mangari area in SE Kenya, *Ore Geology Rev.*, vol. 14, pp. 83–104.

Miller, Donald S., and Stephen Lakatos (1983): "Uplift rate of Adirondack anorthosite measured by fission-track analysis of apatite," *Geology*, vol. 11, no. 5, pp. 284–286.

Milwaukee Sentinel (10 Nov. 1977).

Minning, D.M., A.J. Gow, J. Bonaventura, R. Braun, M. Dewhirst, D.E. Goldberg, and J.S. Stamler (1999): "Ascaris haemoglobin is a nitric oxide-activated 'deoxygenase'," *Nature*, vol. 401, pp. 497–502.

Monchoux P., F. Fontan, P. De Parseval, R.F. Martin, and R.C. Wang (2006): "Igneous albitite dykes in orogenic lherzolite, western Pyrenees, France: A possible source for corundum and alkali feldspar megacrysts in basaltic terranes: 1. Mineralogical association," *Canadian Mineralogist*, vol. 44, pp. 817–842.

Monod: see Hureau and Escudier.
Montesi, Laurent G.J. (28 June 2013): "Solving the mascon mystery," *Science*, vol. 340, pp. 1535–1536.
Moores, E.M. (2002): "Pre-1 Ga (pre-Rodinian) ophiolites: Their tectonic and environmental implications," *Bull. Geological Society of America*, vol. 114, pp. 80–95.
Moralyov, V.M., and M.Z. Glukhovskii (1981): "Circular features of Precambrian shields as recognized on space photographs," *Soviet Journal of Remote Sensing*, no. 3, p. 359.
Muhongo, S., A. Kroener, E. Wallbrecher, P. Tuisku, C. Hauzenberger, and H. Sommer (2001): "Cross-section through the Mozambique Belt of East Africa and implications for Gondwana assembly," *Gondwana Research*, vol. 4, pp. 709–710.
Muller, P.M., and W.L. Sjogren (16 Aug. 1968): "Mascons: Lunar mass concentrations," *Science*, vol. 161, no. 3842, pp. 680–684. DOI: 10.1126/science.161.3842.68.
Müller, Werner E.G., Sergey I. Belikov, Wolfgang Tremel, Carole C. Perry, Winfried W.C. Gieskes, Alexandra Boreiko, and Heinz C. Schroder (2006): "Siliceous spicules in marine demosponges (example *Suberites domuncula*)," *Micron*, vol. 37, pp. 107–120.
Munasinghe, Tissa, and C.B. Dissanayake (1981): "The origin of gemstones of Sri Lanka," *Economic Geology*, vol. 76 (Aug.), pp. 1216–1225.
Murphy, J. Brendan, Gabriel Gutiérrez-Alonso, R. Damian Nance, Javier Fernández-Suárez, J. Duncan Keppie, Cecilio Quesada, Rob A. Strachan, and Jaroslav Dostal (2008): "Tectonic plates come apart at the seams," *American Scientist*, vol. 96, no. 2, pp. 129–137.
Nechaev, V.P., E.V. Nechaeva, A.A. Chashchin, S.V. Vysotskiy, I.T. Graham, and F.L. Sutherland (2009): "New isotopic data on late Cenozoic age and mantle origin of gem zircon and corundum from placers of Primorye, Russia," *Doklady Earth Sciences*, vol. 429A, no. 9, pp. 1426–1429. ISSN 1028_334X.
Nicolini, Pierre (1980): *Méthodes de la gîtologie, ensembles et variances*, Editions Henri Deschamps, Port-au-Prince, Haiti.
Noffke, N., and S.M. Awramik (2013): "Stromatolites and MISS: Differences between relatives," *GSA Today*, vol. 23, no. 9, pp. 4–9. DOI: 10.1130/GSATG187A.1.
Norton, O. Richard (2002): *The Cambridge Encyclopedia of Meteorites*, Cambridge University Press. pp. 121–124. ISBN 0-521-62143-7.
Nursall, J.R. (1959): "Oxygen as a prerequisite to the origin of Metazoa," *Nature*, vol. 183, pp. 1170–1172.
Och, L.M., and G.A. Shields-Zhou (2012): "The Neoproterozoic oxygenation event: Environmental perturbations and biogeochemical cycling.," *Earth-Science Rev.*, vol. 110, no. 1/4, pp. 26–57. DOI: 10.1016/j.earscirev.2011.09.004.
O'Leary, Brian T. (1968): "Influence of lunar mascons on its dynamical figure," *Nature*, vol. 220, no. 5174, p. 1309.
Olivier, Bernard (2006): *The Geology and Petrology of the Merelani Deposit, NE Tanzania*, Ph.D. thesis, University of Stellenbosch.
Opperman, Renée (Jan.–Mar. 1981): "The S[outh] W[estern] A[nthropological] A[ssociation] Symposium 'Astronomy in Anthropology,'" *Archaeoastronomy (The Bulletin of the Center for Archaeoastronomy)*, vol. IV, no. 1, pp. 4–5, College Park, Md.
Osanai, Y., T. Ueno, N. Tsuchiya, Y. Takahashi, Y. Tainosho, and K. Shiraishi (1990): "Finding of vanadium-bearing garnet from the Sør Rondane Mountains, East Antarctica," *Antarctic Rec. (Japan)*, vol. 34, no. 3, pp. 279–291.
Ottaway, T.L., F.J. Wicks, L.T. Bryndzia, T.K. Keyser, and E.T.C. Spooner, (1994): "Formation of the Muzo hydrothermal emerald deposit in Colombia," *Nature*, vol. 369, pp. 552–554.
Panjikar, Jayshree, and Aatish Panjikar (Oct. 2013): "Yellow blue bicoloured sapphires from Chinnadharapuram, South India," *33rd International Gemmological Conference IGC*, (abstract proceedings) Hanoi, Vietnam, pp. 44–47. www.igc-gemmology.org.
Parcha, S.K., B.P. Singh, and Birendra P. Singh (2005): "Palaeoecological significance of the ichnofossils from the Early Cambrian succession of the Spiti Valley, Tethys Himalaya, India," *Current Science*, vol. 88, no. 1, pp. 158–162.

Partington, J.R. (May 1937): "Report of discussion upon chemical and alchemical symbolism: The origins of the planetary symbols for the metals," *Ambix*, vol. 1, no. 1, pp. 61–64.
Pennisi, Elizabeth (15 Feb. 2013a): "Opsins: Not just for eyes," *Science*, vol. 339, pp. 754–755.
Pennisi, Elizabeth (15 Nov. 2013b): "The man who bottled evolution: Richard Lenski's 25-year experiment in bacterial evolution shows no signs of running out of surprises about how mutation and selection shape living things," *Science*, vol. 242, pp. 790–793.
Petroleum History Society, Canada (Nov. 2002): *Newsletter*, vol. XIII, no. 5, http://www.petroleumhistory.ca/archivesnews/2002/phs02-11/phs02-11.pdf.
Physical Map of Asia, http://geology.com/world/asia-physical-map.shtml.
Pin C., P. Monchoux, J.-L. Paquette, B. Azambre, R.C. Wang, and R.F. Martin (2006): Igneous albitite dikes in orogenic lherzolites, western Pyrénées, France: A possible source for corundum and alkali feldspar xenocrysts in basaltic terranes, II. Geochemical and petrogenetic considerations, *Canadian Mineralogist*, vol. 44, pp. 837–850.
Piqué, A., M. Dahmani, D. Jeanette, and L. Bahi (1987): "Permanence of structural lines in Morocco from Precambrian to Present," *Jour. African Earth Sciences*, vol. 6, no. 3 pp. 247–256.
Plotnikova, Irina N. (2006): "Nonconventional hydrocarbon targets in the crystalline basement, and the problem of the recent replenishment of hydrocarbon reserves," *Journal of Geochemical Exploration*, vol. 89, pp. 335–338.
Plotnikova, Irina N. (2008): "New data on the present-day active fluid regime of fractured zones of crystalline basement and sedimentary cover in the eastern part of the Volga-Ural region," *International Journal of Earth Sciences (Geol. Rundschau)*, vol. 97, pp. 1131–1142.
Plotnikova, Irina N., Fidania F. Nosova, and Nikita V. Pronin (in preparation, 2013): "Kerogen of Precambrian crystalline rocks of the Volga-Ural region with emphasis on deep sources of oil deposits," presented at the *Kazan workshop on abiotic hydrocarbons, 13–17 April 2013*, Kazan Federal University, Kazan.
Porter, Susannah M. (2007): "Seawater chemistry and early carbonate biomineralization," *Science*, vol. 316, pp. 1302–1304.
Powers, Sidney (Aug. 1932): "Notes on minor occurrences of oil, gas, and bitumen with various igneous and metamorphic rocks," *Bull. American Association of Petroleum Geologists*, vol. 16, no. 8, pp. 837–858.
Powers, Sidney, and Frederick G. Clapp (Aug. 1932): "Nature and origin of occurrences of oil, gas, and bitumen in igneous and metamorphic rocks," *Bull. American Association of Petroleum Geologists*, vol. 16, no. 8, pp. 719–726.
Pulfrey, William; see Thompson and Dodson (1958).
Raymond, Jason, and Daniel Segrè (24 March 2006): "The effect of oxygen on biochemical networks and the evolution of complex life," *Science*, vol. 311, pp. 1764–1767. DOI: 10.1126/science.1118439.
Revel, J.-P. (1988): "The oldest multicelluar animal and its junctions," In: *Gap Junctions*, Modern Cell Biology, vol. 7, pp. 135–149, E.L. Herztberg, and R.G. Johnson (eds.), Alan R. Liss, New York.
Ridgwell, A.J., M.J. Kennedy, and K. Caldeira (2003): "Carbonate deposition, climate stability, and Neoproterozoic ice ages," *Science*, vol. 302, pp. 859–862.
Robert, Francis (10 Aug. 2001): "The origin of water on Earth," *Science*, vol. 393, no. 5532, pp. 1056–1058. DOI: 10.1126/science.1064051.
Robertson, A.D.C., and F.L. Sutherland (1992): "Possible ages and origins of sapphire and diamond from the Central Queensland gem fields," *Rec. Australian Museum*, Supplement 15, pp. 45–54.
Robinson, Robert (1963): "Duplex origin of petroleum," *Nature*, vol. 199, no. 4889, pp. 113–114. DOI: 10.1038/199113a0.

Rogov, Vladimir, Vasiliy Marusin, Natalia Bykova, Yuriy Goy, Konstantin Nagovitsin, Boris Kochnev, Galina Karlova, and Dmitry Grazhdankin (2012): "The oldest evidence of bioturbation on Earth," *Geology*, vol. 40, no. 5, pp. 395–398.

Rondeau, Benjamin, and Emmanuel Fritsch (2013): "Les gemmes de l'Afrique de l'Est: ce qui est Néoprotérozoïque… ou ce qui ne l'est pas!" *Règne Minéral,* Cahier 2 - 2013, pp. 129–141.

Routhier, Pierre (1968): "Sur trois principes généraux de la metallogénie et de la recherche minérale," *Mineralium Deposita* (Berlin) vol. 4, p. 217.

Royal Geographical Society, *Proceedings,* opening session (1884–1885).

Russell, James K., Lucy A. Porritt, Yan Lavallée, and Donald B. Dingwell (2012): "Kimberlite ascent by assimilation-fuelled buoyancy," *Nature*, vol. 481, pp. 352–356. DOI: 10.1038/nature10740.

Ryder, Graham, Christian Koeberl, and Stephen J. Mojzsis (2000): "Heavy bombardment of the Earth at ~3.85 Ga: The search for petrographic and geochemical evidence," p. 480, In: Robin Canup and Kevin Righter, eds., *Origin of the Earth and Moon,* University of Arizona Press, Tucson, pp. 475–492.

Sander, Bruno (1948): *Einführung in die Gefügekunde der Geolischen Körper*, vol. 1, Springer, Vienna.

Sanders, L.D. (1965): "Geology of the contact between the Nyanza Shield and the Mozambique Belt in Western Kenya," *Bull. Geological Survey of Kenya*, no. 7.

Sankaran, A.V. (10 Jan. 2007): "Closure of the East-West Gondwana divide: The dawn of 'Cambrian Explosion,'" *Current Science,* vol. 92, no. 1, pp. 19–20.

Saul, John M. (1970): "Gemstones in Gondwanaland," *EOS, Trans. American Geophysical Union,* vol. 51, no. 11 (Abstract T19), p. 825.

Saul, John M. (1977): "Large circles on the Earth's surface," *Meteoritics*, vol. 12, no. 3 (abstract), pp. 358–359.

Saul, John M. (1978): "Circular structures of large scale and great age on the Earth's surface," *Nature*, vol. 271, no. 5643, pp. 345–349.

Saul, John M. (1981): "The origin of diamonds and the Deep Gas Hypothesis," XVIII International Gemmological Conference, proceedings, *Jour. Gemmological Society Japan*, vol. 8, no. 1–4, pp. 79–85. http://phe.rockefeller.edu/news/archives/1999

Saul, John M. (1984): "Deposits of coloured gemstones: Their distribution in time and space," *Jour. Gemmologists Association of Sri Lanka,* no. 1, pp. 13–16.

Saul, John M. (1985): "The geological distribution of coloured gemstones," *20th International Gemmological Conference* (abstract), Sydney, Australia.

Saul, John M. (Jan. 1994): "Cancer and autoimmune disease: A Cambrian couple?" *Geology*, vol. 22, no. 1, p. 5.

Saul, John M. (2006): "'Ice meteorites': Is it prudent to ignore anecdotal reports?" *Meteorite*, vol. 12, no. 2, pp. 20–21.

Saul, John M. (2007a): "Origin of crystalline colored gemstones in Gondwanaland and the Greater Himalayan Region," *30th International Gemmological Conference* (abstract) Moscow, pp. 87–89.

Saul, John M. (2007b): "Origin of the phyla and cancer," *Lethaia,* vol. 40, no. 4, pp. 359–363. DOI: 10.1111/j.1502-3931.2007.00034.x.

Saul, John M. (2009): "Did detoxification processes cause complex life to emerge?" *Lethaia*, vol. 42, pp. 179–184. DOI: 10.1111/j.1502-3931.2008.00126.x.

Saul, John M. (2013): *The Tale Told in All Lands*, Les 3 Colonnes, Paris. ISBN: 978-2-84608-350-8. http://www.tomebook2.com.

Saul, John M., and Janice A. Glaholm (1981): "Is Lake Victoria a crater lake?" *Fourth International Conference of Basement Tectonics*, Abstract volume, Oslo, Aug. 10–14, 1981.

Saul, John M., and Laurent Schwartz (2007): "Cancer as a consequence of the rising level of oxygen in the Late Precambrian," *Lethaia*, vol. 40, no. 3, pp. 211–220. DOI: 10.1111/j.1502-3931.2007.00014.x.

Sawyer, Edward W., Bernado Cesare, and Michael Brown (Aug. 2011): "When the continental crust melts," *Elements*, vol. 7, pp. 229–234.
Schmetzer, Karl, and Anna-Kathrin Malsy (2011): "Alexandrite and colour-change chrysoberyl from the Lake Manyara alexandrite-emerald deposit in northern Tanzania," Jour. Gemmology, vol. 32, no. 5–8, pp. 179–209.
Schmidt, A., and K. Mack (1913): Württ. Jahrbücher f. Statist. u. Landleskde., Jahrg. 1912, Heft I, p. 96 et seq., as cited by Gold and Soter (1979).
Schultz, P.H., and R.A. Schultz (1980): "Ancient impact basins on Mars," In: *Multi-ring Basins: Formation and Evolution* (abstract), pp. 77–79. Lunar and Planetary Institute Contribution 414, R.B. Merrill, and P.H. Schultz, eds.
Schwandner, Florian M., Terry M. Seward, Andrew P. Giże, Keith Hall, and Volker J. Dietrich (2013): "Halocarbons and other trace heteroatomic organic compounds in volcanic gases from Vulcano (Aeolian Islands, Italy)," *Geochimica et Cosmochimica Acta*, vol. 101, pp. 191–221.
Schwartz, Laurent (2004): *Cancer: Between Glycolysis and Physical Constraint*, Springer, Berlin-Heidelberg, 150 pp.
Schwartz, Laurent, J. Balosso, F. Baillet, B. Brun, J.-P. Amman, and A.J. Sasco (2002): "Cancer: The role of extracellular disease," *Medical Hypotheses*, vol. 58, no. 4, pp. 340–346.
Schwarz, Dietmar, Gaston Giuliani, Virginie Garnier, and Daniel Ohnenstetter (2007): "Marble-type ruby deposits: A review," *30th International Gemmological Conference* (abstract) Moscow.
Science (4 Nov. 2011; p. 587); (15 June 2012; p. 1366).
Science News (29 April 1978; p.281); (28 Aug. 1999; p. 141); (8 Nov. 2003; pp. 293–294).
Scott, R.A., J.P. Howard, L. Guo, R. Schekoldin, and V. Pease (2010): "Offset and curvature of the Novaya Zemlya fold-and-thrust belt, Arctic Russia," *Geological Society, London, Petroleum Geology Conference Series*, vol. 7, pp. 645–657. DOI: 10.1144/0070645.
Seilacher, Adolf (1983): "Late Precambrian and Early Cambrian Metazoa; preservational or real extinctions?" In: *Patterns of Change in Earth Evolution*, pp.159–68, H.D. Holland and A.F. Trendall, eds., Berlin. ISBN 978-3-642-69319-9.
Seilacher, Adolf (1989): "Vendozoa: Organismic construction in the Proterozoic biosphere," *Lethaia*, vol. 22, pp. 229–39.
Seilacher, Adolf (1992): "Vendobionta and Psammocorallia—lost constructions of Precambrian evolution," *Jour. Geol. Soc. London*, vol. 149, pp. 607–613.
Seilacher, Adolf (2007): "The nature of vendobionts," In: *The Rise and Fall of the Ediacaran Biota*, P. Vickers-Rich and P. Komarower, eds., Geological Society Special Publication no. 286, The Geological Society (London), pp. 387–398.
Seilacher, A., D. Grazhdankin, and A. Legouta (2003): "Ediacaran biota: The dawn of animal life in the shadow of giant protists," *Paleontological Research*, vol. 7, pp. 43–54.
Sheridan, J.D. (1987): "Cell communication and growth," In: *Cell to Cell Communication*, W.C. DeMello, ed., pp. 187–222, Plenum Press, New York.
Shields-Zhou, G.A., and L.M. Och (2011): "The case for a Neoproterozoic oxygenation event: Geochemical evidence and biological consequences," *GSA Today*, vol. 21, no. 3, pp. 4–11.
Shirley, James H., and Rhodes W. Fairbridge (1997): *Encyclopedia of Planetary Sciences*, Springer, p. 196.
Signor, Philip W., and Jere H. Lipps (1992): "Origin and early radiation of the Metazoa," In: J.H. Lipps and P.W. Signor, eds., *Origin and Early Evolution of the Metazoa*, pp. 3–23, Plenum, New York.
Signor, Philip W., and Dallas A. Ryan (1993): "Lower Cambrian fossil *Volborthella*: The whole truth or just a piece of the beast?" *Geology*, vol. 21, no. 9, pp. 805–808.
Silver, Paul D., and Mark D. Behn (2008): "Intermittent plate tectonics?" *Science*, vol. 319, pp. 85–88.

Silver, Paul D., Mark D. Behn, Katherine Kelley, Mark Schmitz, and Brian Savage (2006): "Understanding cratonic flood basalts," *Earth and Planetary Science Letters,* vol. 245, pp. 190–201.

Simkiss, K. (1977): "Biomineralization and detoxification," *Calcified Tissue Research,* vol. 24, no.1, pp. 199–200.

Simkiss, K., and K.M. Wilbur (1989): *Biomineralization,* 337 pp., Academic Press, San Diego, Calif.

Simmons, W.B. "Skip" (2007): "Gem-bearing pegmatites," In: *Geology of Gem Deposits,* Lee A. Groat, ed., Mineralogical Association of Canada, Short Course Series, vol. 37, pp. 169–206, Yellowknife, N.W.T.

Simmons, W.B., and K.L. Webber (2008): "Pegmatite genesis: State of the art," *European Jour. Mineralogy,* vol. 20, no. 4, pp. 421–438.

Sonnenschein, Carlos, and Ana Soto (1999): *The Society of Cells,* 154pp., Bios Scientific, Oxford, UK.

Soter, Steven (1999): "Macroscopic seismic anomalies and submarine pockmarks in the Corinth–Patras rift, Greece," *Tectonophysics,* vol. 308, pp. 275–290.

Squire, Richard J., Ian H. Campbell, Charlotte M. Allen, and Christopher J.L. Wilson (2006): "Did the Transgondwanan Supermountain trigger the explosive radiation of animals on Earth?" *Earth and Planetary Science Letters,* vol. 250, pp. 116–133.

Stern, Robert J. (1994): "Arc assembly and continental collisions in the Neoproterozoic East African orogen: Implications for the consolidation of Gondwanaland," *Ann. Rev. Earth and Planetary Sciences*, vol. 22, pp. 319–351.

Stern, Robert J. (2005): "Evidence from ophiolites, blueschists, and ultrahigh-pressure metamorphic terranes that the modern episode of subduction tectonics began in Neoproterozoic time," *Geology,* vol. 33, no. 7, pp. 557–560.

Stern, Robert J. (Mar. 2007): "When and how did plate tectonics begin? Theoretical and empirical considerations," *Chinese Science Bull.*, vol. 52, no. 5, pp. 578–591. DOI: 10.1007/s11434-007-0073-8.

Stern, Robert J. (2008): Neoproterozoic crustal growth: The solid Earth system during a critical episode of Earth History, *Gondwana Research*, vol. 14, pp. 33–50.

Stern, Robert J., Tatsuki Tsujimori, George Harlow, and Lee A. Groat (2013): "Plate tectonic gemstones," *Geology,* vol. 41, no. 7, pp. 723–726, with supplementary material from Data Repository item 2013202.

Sunday Telegraph, London (17 Jan. 1999).

Sutherland, F.L. (1999): "Volcanism, geotherms, gemstones and lithosphere, since orogenesis, N.E. New South Wales: A synthesis," In: P.G. Flood, ed., *New England Orogen: Regional Geology Tectonics and Metallogenesis*, Earth Sciences, University of New England, Armidale, pp. 354–364.

Sutherland, F.L., G. Bosshart, C.M. Fanning, P.W.O. Hoskin, R.R. Coenraads (2002): "Sapphire crystallization, age and origin, Ban Huai Sai, Laos: Age based on zircon inclusions," *Jour. Asian Earth Sciences*, vol. 20, pp. 841–849.

Suwa, K., K. Suzuki, K. Miyakawa, and T. Agata (1979): "Vanadian and vanadium grossular from Mozambique metamorphic rocks, Mgama Ridge, Kenya," *4th Preliminary Report on African Studies,* Nagoya University, pp. 87–96.

Taylor, Dan, Andrew H. Rankin, and Peter J. Treloar (2013): "Liquid hydrogen sulphide (H_2S) fluid inclusions in unheated tanzanites (zoisite) from Merelani, Tanzania: Part 1. Recognition, characterization and gemological importance," *Jour. of Gemmology,* vol. 33, nos. 5–6, pp. 149–159.

Taylor, Matthew C. (2005): "The sol-gel nature of pegmatites," *Elba 2005 - International Meeting on Crystallization Processes in Granitic Pegmatites, Abstracts*, Elba, Italy, May 23–30. http://www.elba-pegmatites.org/Abstracts.doc.

Taylor, Matthew C. (2006): "The gel model for the formation of gem-bearing pockets within granitic pegmatites, and implications for gem synthesis," *Gems & Gemology*, vol. 42, no. 3, pp. 110–111.

Taylor, S.R. (1993): "Early accretion history of the Earth and the Moon-forming event," *Lithos*, vol. 30, pp. 207–221.

Tectonic Map of North America, 1:5,000,000: (1969) compiled by Philip B. King, U.S. Geological Survey, Washington, D.C.

Themelis, Ted (2008): *Gems and Mines of Mogok*, published by the author in Thailand, 352 pp. ISBN 0940965-30-5.

Thompson, A.O., and R.G. Dodson (1958): "Geology of the Derkali-Melka Murri Area," *Report, Geological Survey Kenya*, no. 43.

Thompson, Helen (26 Apr. 2012): "Ancient asteroids kept on coming," *Nature*, vol. 484, p. 429.

Touret, Jacques L.R., and Jan Marten Huizenga (2012): "Fluid-assisted granulite metamorphism: A continental journey," *Gondwana Research*, vol. 21, pp. 224–235. http://www.sciencedirect.com/science/article/pii/S1342937X11002140.

Touret, Lydie (Mar. 1992): "Les pegmatites de Volhynie (Ukraine)," *Revue de Gemmologie, a.f.g.*, Association Française de Gemmologie, Paris no. 110, pp. 25–26.

Touret, Lydie (2010): "Digne de Jules Verne: Jean-Pierre Alibert (1820–1905)," In: *Notre Terre, ce joyau*, pp. 70–85, L. Touret, ed., Musée de minéralogie / Mines-ParisTech, Paris.

Towe, K.M. (1970): "Oxygen-collagen priority and the early metazoan fossil record," *Proceedings of the National Academy of Sciences, U.S.A.*, vol. 65, pp. 781–788.

Towe, K.M. (1981): "Biochemical keys to the emergence of complex life," In: J. Billingham, ed., *Life in the Universe*, pp. 297–305. MIT Press, Cambridge, Mass.

Tributsch, Helmut (1982): *When Snakes Awake: Animals and Earthquake Prediction,* MIT Press, Cambridge, Mass.

Trofimov, Vladimir A. (2013): "Oil-refilling channels: forecasting, activity estimation, technology of localization and drilling-in," *Kazan workshop on abiotic hydrocarbons, 13–17 April 2013* (abstract) I.N. Plotnikova and A.S. Borisov, eds., Kazan Federal University, Kazan, pp. 58–60. ISBN 978-5-00019-053-1.

Trosko, James, E. (Aug. 2007): "Gap junctional intercellular communication as a biological 'Rosetta Stone' in understanding, in a systems biological manner, stem cell behavior, mechanisms of epigenetic toxicology, chemoprevention and chemotherapy," *Journal of Membrane Biology,* vol. 218, nos. 1–3, pp. 93–100. DOI 10.1007/s00232-007-9072-6.

Trosko, James, E. (March 2011): "The gap junction as a 'Biological Rosetta Stone': Implications of evolution, stem cells to homeostatic regulation of health and disease in the Barker hypothesis," *Journal of Cell Communication and Signalings*, vol. 5, no. 1, pp. 53–66.

Trujillo, C.A., M.E. Brown, K.M. Barkume, E.L. Schaller, and D.L. Rabinowitz (2007): "The surface of 3003 EL61 in the near infrared," *Astronomical Journal - Astrophysics,* vol. 655, pp. 1172–1178.

Tsunogae, T., and M. Santosh (2010): "Ultrahigh-temperature metamorphism and decompression history of sapphirine granulites from Rajapalayam, Southern India: Implications for the formation of hot orogens during Gondwana assembly," *Geological Magazine,* vol. 147, pp. 42–58.

Vali, H., and Kirschvink, J.L. (1990): "Observation of magnetosome organization, surface structure and iron biomineralization of undescribed magnetic bacteria: Evolutionary speculations," In: R.P. Blakemore and R.B. Frankel, eds., *Iron Biomineralization*, pp. 97–115, Plenum Publishing Corporation, New York.

Veevers, J.J. (2001): *Atlas of Billion-year earth history of Australia and neighbours in Gondwanaland*, GEMOC Press, Sydney.

Veevers, J.J. (2003): "Pan-African is Pan-Gondwanaland: Oblique convergence drives rotation during 650–500 Ma assembly," *Geology*, vol. 31, no. 6, pp. 501–504.

Vitanage, P.W. (1972): "Post-Precambrian uplifts and regional tectonic movements in Ceylon," *24th International Geological Congress*, Sec. 3, pp. 642–654, Montreal.

Von Dechend, Hertha (1961, 1966, 1979): unpublished lecture notes, Massachusetts Institute of Technology, Cambridge, Mass.

Wakita, H., Y. Nakimura, K. Notsu, M. Noguchi, and T. Asada. (1980): "Radon anomaly: A possible precursor of the 1978 Izu-Oshima-kinkai earthquake," *Science*, vol. 207, pp. 882–883.

Wall Street Journal (8 June 1977; 16 Apr. 1999).

Walton, Lori (2004): "Exploration criteria for coloured gemstone deposits in the Yukon," Yukon Energy, Open File no. 2010-10, Mines and Resources, Whitehorse, Yuk.

Walton, Michael T. (July 1976): "John Dee's *Monas Hieroglyphica*: Geometric Cabala," *Ambix*, vol. XXIII, no. 2, pp. 116–123.

Ward, Peter D. (2006): *Out of Thin Air: Dinosaurs, Birds, and Earth's Ancient Atmosphere*, Joseph Henry Press, Washington, D.C.

Weatherley, Dion K., and Richard W. Henley (2013): "Flash vaporization during earthquakes evidenced by gold deposits," *Nature Geoscience*, vol. 6, pp. 294–298. DOI: 10.1038/ngeo1759.

Webber, K.L., W.B. Simmons, A.U. Falster, and E.E. Foord (1999): "Cooling rates and crystallization dynamics of shallow level pegmatite-aplite dikes, San Diego County, California," *American Mineralogist*, vol. 84, pp. 708–717.

Whitaker, E.A. (1981 [1980]): "The lunar Procellarum basin," In: *Multi-ring Basins: Formation and Evolution*; Proc. Lunar and Planetary Science Conference (Nov. 10–12, 1980), pp. 105–111, Houston, Pergamon Press, New York and Oxford.

Wicke, Charles R. (1984): "The Mesoamerican rabbit in the Moon: An influence from Han China?" *Archaeoastronomy (The Bulletin of the Center for Archaeoastronomy)*, vol. VII, nos. 1–4, pp. 46–55, College Park, Md.

Wieczorek M., G.A. Neumann, F. Nimmo, W. Kiefer, G. Taylor, H. Melosh, R. Phillips, S. Solomon, J. Andrews-Hanna, S. Asmar, A. Konopliv, F. Lemoine, D. Smith, M. Watkins, J. Williams, M. Zuber (8 Feb. 2013): "The crust of the Moon as seen by GRAIL," *Science*, vol. 339, pp. 671–675. DOI: 10.1126/science.1231530.

Wiens, Douglas A., and Nathaniel O. Snider (24 Aug. 2001): "Repeating deep earthquakes: Evidence for fault reactivation at great depth," *Science*, vol. 293, pp. 1463–1467.

Xiao, Shuhai, and Marc Laflamme (2009): "On the eve of animal radiation: phylogeny, ecology and evolution of the Ediacara biota," *Trends in Ecology & Evolution*, vol. 24, no. 1, pp. 31–40.

Yakubchuk, Alexander (2008): "The gyroscopic Earth and its role in supercontinent and metallogenic cycles," *Ore Geology Rev.*, vol. 34, pp. 387–398.

Yates, Frances A. (1978): *Giordano Bruno and the Hermetic Tradition*, Routledge and Kegan Paul, London.

Yui, T.F., C.-M. Wu, P. Limkatrun, W. Sricharn, and A. Boonsoong (2006): "Oxygen isotope studies on placer sapphire and ruby in the Chanthaburi-Trat alkali basaltic gemfield, Thailand," *Lithos*, vol. 86, pp. 197–211.

Zaw, Khin, F.L. Sutherland, F. Dellapasqua, C.G. Ryan, Tzen-Fu Yui, T.P. Mernagh, and D. Duncan (Dec. 2006): "Contrasts in gem corundum characteristics, eastern Australian basaltic fields: Trace elements, fluid/melt inclusions and oxygen isotopes," *Mineralogical Magazine*, vol. 70, no. 6, pp. 669–687. DOI: 10.1180/0026461067060356.

Zubkov, V.S. (2009): "Hypotheses of heavy hydrocarbons and bitumen genesis in different-age ophiolites," *Litosphera*, vol. 1, pp. 70–80 (in Russian).

Zwaan, J.C. (2006): "Gemmology, geology and origin of the Sandawana emerald deposits, Zimbabwe," *Scripta Geologica*, vol. 131, pp. 1–211. https://science.naturalis.nl/en/people/scientists/hanco-zwaan/.

About the Author

The author was born in New York City in 1937. He worked as an undergraduate geologist in West Africa. On receiving a Ph.D. in geology from the Massachusetts Institute of Technology in 1964, he moved to East Africa and then to Paris. Working in the area east of Mount Kenya in the late 1960s, he discovered and mined a number of small deposits of beryl, an ore of the metal beryllium. Some crystals of beryl turned out to be sufficiently transparent to be faceted as the gemstone aquamarine, and this led to a continuing interest in gems and gemology. He subsequently discovered major deposits of gemstones, including what has become known as the John Saul Ruby Mine, and has received awards and honors from several professional gemological groups. Throughout his career he has followed two other subjects that have intrigued him since his student days, meteorite impact craters and the sudden appearance in the fossil record of the first complex animals, an episode commonly referred to as the Cambrian Explosion. His publications have appeared in *Nature*, *The Journal of Geophysical Research*, *Gems & Gemology*, *Ore Geology Reviews*, and journals in Australia, France, Germany, Japan, Nigeria, and Sri Lanka. He recently published *The Tale Told in All Lands* (Les Trois Colonnes, Paris, 2013, 623 pages).

Imprimé en France
ISBN 978-2-37081-004-5
Dépôt légal : 1er trimestre 2014